BEYOND THE SUBJECTIVITY TRAP

MARTIN O'DEA

imprint-academic.com

Copyright © Martin O'Dea, 2015

The moral rights of the author have been asserted.
No part of this publication may be reproduced in any form
without permission, except for the quotation of brief passages
in criticism and discussion.

Published in the UK by
Imprint Academic, PO Box 200, Exeter EX5 5YX, UK

Distributed in the USA by
Ingram Book Company,
One Ingram Blvd., La Vergne, TN 37086, USA

ISBN 9781845407858

A CIP catalogue record for this book is available from the
British Library and US Library of Congress

Contents

Prologue	v
Book One	
Chapter One. Consistent Hidden Revolutions	1
Chapter Two. The World Around Us	5
Chapter Three. What is Stuff Made of?	21
Chapter Four. All About Us	24
Book Two	
Chapter Five. What is the Subjectivity Trap?	37
Chapter Six. Ways to See Beyond the Subjectivity Trap	70
Chapter Seven. Dearest	79
Chapter Eight. Why is this Such a Hard Sell?	82
Interlude. What is Technology Anyway?	112
Book Three	
Chapter Nine. Beyond the Subjectivity Trap	125

Prologue

There are some issues we encounter where we debate and argue different sides, become involved emotionally, and perhaps refine our position over time—we would never say in truth that we are absolutely and unwaveringly certain on these issues. There are other issues where, when practicable, we would state with confidence that we are as certain as we can be—or, at the very least, we are not going to return to our earlier position. One might look at gaining understanding of the spherical nature of the planet, or that the moon does not walk specifically with an individual or that certain fairytales and stories adults tell us when we are kids are just that. These are instances where we at some point put away such 'childish thoughts' not to return.

This is how I feel towards some issues that I wish to discuss here. My understanding of consciousness and thought is that these are mere functional biological processes that can be comprehended entirely through their physical manifestations, and are no more special than a dog's bark, an ant's progress, or the growth of a tree, or the orbit of a rock! These may seem big claims (or at least they are potentially controversial for a lot of readers, and require a lot of support)—hence, this book.

There will be minimal scientific content or mathematical concepts used in this book. There may be sections that become convoluted and complex conceptually. My intention is to express the key points as succinctly as possible; this may not be the result, though, such is the topic under consideration. However, there is so much great stuff written already and digestible, by anyone who is curious—no need to be a specialist or in a relevant field—that I would suggest you use the wonders of the internet to read and watch talks, lectures, and debates in any area I mention that may not be very familiar.

Book One

Chapter One

Consistent Hidden Revolutions

There have been a number of well-known instances through history where large swathes of the then widely held worldview, or what was 'known' by humanity, were completely revolutionised and largely dismissed. There is, of course, a natural accumulation of knowledge also, but in many cases the context in which things were understood was turned on its head. The flat world and the Sun orbiting the Earth (which is not at the centre of the universe), the absoluteness of time, or the influence of cleanliness on illness were all major shifts that took time to move from discovery and proof to widespread acceptance. In each instance the ideas were resisted and seen as heretical—they were resisted so strongly that from this perspective it might be worth pondering why new ideas of large significance meet such strong resistance—and this, in fact, is something that will be discussed later on when we take a look at modern theories and understanding of mind; and, particularly, at certainty.

When Albert Einstein became world famous and the 'masses' tried desperately through a plethora of analogies to get to grips with the ideas of relativity, a gap perhaps first emerged between what the students of fact or scientific seekers of truth were presenting and the general public's ability to absorb what this meant for them. It could be argued that through the next century that gap has been repeatedly extended and, unfortunately, seems to not be bridged too often. There are discoveries and theories abroad now that are revolutionising our understanding of ourselves and our place in the universe that are much more significant than anything from before, the ramifications are much more far reaching, but the changes that are required from our traditional worldviews are greater also.

The hope of this book is to present and discuss some of the fundamental understandings that underpin the increasingly rapid development of new possibilities; to explore latest views of the world around us and within us. To begin to question the concept of thought, itself, and to look at the whole range of incredible repercussions of what we now know and are doing and trying to do, most having the potential for great improvements and great tragedies for humanity.

The genuine hope is that you may at times see the world from a different angle, though very little of what you know or hold dear need change, but the context in which you know these things will. While we may appreciate relativity and quantum mechanics we do not factor them in when driving a car. Nor indeed do we feel queasy at the thought of a spherical world as we walk towards the horizon. So it should be if we can dismiss the primacy of thought!

To get the most from this journey you just need to park as much as possible your current understandings of the world you find yourself in, and what you see to be 'real' or certain. Don't fret, they are retrievable… Read devoid of the bias of all the impossibilities and limits you always thought certain, assess the implications and possibilities that our new understandings point towards—all with nothing but your ability to reason.

It is understandable through studying evolutionary biological terms (as is almost everything else—and this too will be revisited) that our internal world representations are central to our sense of 'self'. You may already be aware, for example, that we see things in 2D (like a TV) and that we *interpret* depth, but this interpretation gains in significance as our ancestors approached the edge of a sheer drop or estimated how far off a beast sits. Our worldview is based on certainties in many such instances. Furthermore, we can really only approach something new with our current worldview and its inherent prejudices, and it is incredibly difficult for us to accept major disagreements with our interpretation of reality.

To change these interpretations, in fact, we need to rewire our brains and how they process images. As we will also see later this requires the things you would think, like education, concentration, and consideration; but it also involves repeat exposures. It is, as we will see, a physical alteration when we 'change' our minds. We need to endure some reality-shifting and therefore baffling scientific findings a number of times before we begin to enjoy them and eventually accept them. Sometimes they counter our intuition so much that no amount of exposure seems to do the trick! It will most likely sound weird but many times as I read an account of relativity or the dual slit experiment in quantum mechanics, I got a real kick out of the fact that my brain tried desperately to find explanations other than those provided by the leading physicists of the last hundred years—just to protect my understanding of 'how things are to me'. There I was in my first few books picking holes in Einstein's and Bohr's work! Eventually, though, we can agree that something is credible, if difficult in some sense to accept. We can, at least, see why other ideas do not stand up by comparison. The thing to undermine is not the specific breakthrough or the fact that we believed hand on heart in something a year ago that now we don't—we must attempt to understand the reasons for and the limitations of our certainties.

We should, though, be confident as amateurs that we can appreciate anything with enough exposure and explanation. It is a worry that children are told in their early school years that they are weak at something and then do not bother to exercise their neural networks in this general field, and so we get a self-fulfilling prophecy. It is essential to keep in mind the value of our own reasoning ability and to not be afraid to disagree. The only thing to be avoided, as history shows, is certainty. Perhaps an example of human reasoning winning out over experience lies in the oft-cited flat vs. spherical world debate. Eratosthenes, a Greek scholar who lived 275–194 BC, heard that on a summer solstice day the Sun is reflected perfectly in a deep well in

Syene. But at the same time and same day, in Alexandria, the sun wasn't reflected perfectly in the same type of well. Why? It occurred to him that the only way that could have happened is by the curvature of the Earth. Easy said, I guess, but this is a stunning piece of deduction. Imagine the Sun directly above a point and no shadow of a stick in the ground being cast at all, yet elsewhere the shadow can be seen.

It is important to bear in mind that there was absolutely no reason for him to imagine that there was anything other than a flat world and a whole range of very strong reasons (not falling off the edge, for one) to see it as everyone else did. So the trigonometry was impressive but the thinking was really amazing, and the benefits of questioning assumptions through deductions are obvious.

With that healthy dose of uncertainty promotion, I want to present a run-through of some of the understandings of the world around us that have been bequeathed to us.

Chapter Two

The World Around Us

THE BIG

The first task, I think, in trying to get a good understanding of our environment is to try to get an idea of size—in truth this is an impracticable task but still fun to try. We'll begin by concentrating on the big, looking up to the skies and considering what we know of what lies out there. There can be a tendency to demur to the past and what was known before (there is nothing new under the sky), etc. However, it is important to bear in mind that four lifetimes ago most people would have imagined that if you could see beyond the clouds you would see heaven directly—and even for those who trained the first telescopes upwards there may well have been the possibility of encountering a big, soft, gentle old man's gigantic eyeball. Two lifetimes ago we could not leave the ground, and talk of someone or something going into space was absolutely crazy. Just one lifetime ago life (as in green men) on Mars would have been likely or possible for most, and the idea of automated computation was practically unknown. Of course we should appreciate and learn from history but sometimes it can be a little self-defeating to allow ourselves to be indebted entirely to its proclamations! When we saw beyond those limits, though, surely none could have envisioned the dimensions?

I imagine we all have reasonable ideas that we are the third planet in a solar system of nine or perhaps eight (it seems that Pluto has been demoted recently). We mostly frame our notion of the size of the solar system from representations we encounter in school. In your mind's eye—I will ask you to activate that an awful lot from now on—we see the diameters of these balls as being not much smaller than the space between each one. In reality, though, we need to consider the following (this is a very well-known thought experiment):

You will need some items to represent the planets, which you should lay out on a big table. A football, a couple of coffee beans, and a few tiny pinheads; these (coffee beans and pinheads) will, of course, represent the planets (massive and small), and the comparatively truly massive football represents the Sun. To complete this thought experiment on scale you now need to forget all about the ruse that was the 'big table'!

To get from the Sun to Mercury you will need to put the football down and take ten full paces (yards) and put down a pinhead representing Mercury, 26 more get you to Earth and another 40 puts you out to Mars; however, you then have to travel the guts of a football field to make your way out to Uranus, and the distances grow from there. In fact, when you have placed the pinhead down for Pluto you will be considerably more than half a mile from the football which you will have lost sight of an awful long way back. Look at the pinhead, look how small it is compared to an inch or a foot around it; now look at the half a mile—this is a nice starting point towards appreciating the vastness of our solar system. Perhaps reassuring when you hear of an asteroid in the solar system and the old back-of-the-school-classroom model arouses fear.

That is our solar system in scale, eight or nine planets orbiting a star, and we know the Earth is an almost unimaginably big thing for us to comprehend or certainly to scale down to a pinhead. We should move out a little bit in our perspectives. This is one thought process that can be great fun. Consider Usain Bolt sprinting past you at nearly 30 miles per hour, amazing stuff how fast he moves. Now change up a gear to a Ferrari out on the open road knocking around 200 miles per hour; step up inside a plane and that amazing feeling where you can see yourself cross the geography of a country, difficult to get a handle on these motions but we are looking at around 500 miles per hour. Well, a quick look will tell you that US military fighter jets can fly in the thousands of miles per hour and Helios 2, a manmade sun probe, clocked 150,000 miles per hour. Light goes at 300,000 miles — per second.

To attempt *in vain* to get a handle on that type of concept try the following — mentally move across your own country in a fast moving plane. Let yourself appreciate however many hours this will take, strike out into neighbouring countries or across vast oceans. Eventually imagine replicating Magellan or Phileas Fogg and circumnavigating the world (a notion, as we discussed, that did not even exist in our earlier flat world model, which survived practically all of humanity's tenure, and then became an obscene dream to occur in a lifetime). OK, what about something zipping around this massive expanse — fields and ditches, rivers and lakes, out into seeming interminable oceans, only to begin again in much, much more massive landscapes, the whole way around, not once, not twice, not thrice, but, in fact, seven times in one… second! If you *can* imagine that then you need to go back and read the last sections again! That is the speed of light. It is totally incomprehensible!

Light travelling at that speed still takes eight minutes, or 480 of those seconds-based trips, to get from the Sun to Earth. Imagine something going so fast that it gets around the world seven times in a second, bursting out into space, now it moves those vast distances repeatedly each second for eight whole minutes. If you don't already know then you can guess that a light year is how far this light would have traversed at the end of one year. Nuts? OK, but our nearest star (other than the Sun) — called Proxima Centuri — is four years and four months away. I hope your head is spinning now and you want to give up trying to keep human appreciations of these speeds and distances. Here is your get-out phase if you're still with it — there are approximately 200–400 billion stars in the Milky Way galaxy; and there are somewhere in the region of 200 billion galaxies in the observable universe.

The terrible thing about that last sentence is it lets our attempts to grasp our environment off the hook. Bear in mind the distance to our nearest star, the obscene distance (go around the world seven times in one second at that speed, keep going for four and a half years) — then rethink that big, spotty, creamy spiral cluster we get shown to represent the Milky Way and realise that there is on average around five light years between each star (though many stars come in pairs much closer together). Even having re-evaluated that cluster image and forcing ourselves to see spaces within as essentially vastly empty with incredible differences between each entity, and bearing in mind that the Milky Way would take us (at that seven-times-around-the-world-in-one-second speed) one hundred thousand years (the guts of all human history) to cross — well, we still have to try in our ever less reliable mind's eye to understand that galaxies themselves are chunks of matter in an otherwise empty universe; and, as we said, there are estimates of around 200 billion galaxies in the observable universe.

Why do I feel that lets us off the hook, though? When we say there are 200 billion stars in the Milky Way and 200 billion galaxies in the observable universe we must really pause and go back and say to ourselves, 'so the distance between Earth and the Sun is…' — 'the Milky Way is…' — and so on. And when we have gotten back to the four-plus years to the nearest star, then we try to concentrate on the 200–400 billion stars. Unfortunately, in these debt-ridden global financial times one billion seems small change as a number, but go to the trouble of figuring it out.

We probably should have talked about the vastness of the Sun at this point — the pinhead-to-football and the already discussed Magellan circumnavigation might help — but there is another sun and another — in some arbitrary space there are an immense 100 suns; then another one of those immense collections and another up to a hundred more of those volumes of space, and eventually we have 10,000 suns. Now set about repeating that whole process (all of it) two million times. The key thing then is defeated as our comprehension is — remember when you change multiples, don't forget the actual constituent parts, so when you do that whole exercise again (the one we couldn't really do a while ago); we have two galaxies, and when we go from 190,127 million galaxies to 190,128 million galaxies we have to do the whole process again, each single time one million times. So, space. A big place!

Finally, I used the term earlier 'the observable universe'; this does not mean the total universe—it includes an event horizon—this means that we can only see that which has travelled to us at the speed of light over the last 14 billion years, or since the Big Bang.

It is easy to see why so many cite the vastness of space as evidence of our own insignificance.

P.S. One little side note at this point—when we talk about the vastness of space, we now know that it has been so for many billions of years (of course it appears ever expanding but let's park that for a moment). This is a good example of how thick and fast reality shattering information comes to us. Ponder for a while what an individual (particularly a non-scientist) would have imagined the size of space or the distances to stars or the speed of light to have been just a few centuries ago. I would repeat the argument that looking at history as a guide is vital—but using it as an excuse to repeat falsehoods and make mistakes is unforgivable in that context of 'ah, well, nothing ever really changes'—we do not have the excuses of degrees of relative ignorance most all of mankind have had.

The small

Next we should journey downward from the everyday. Democritus coined the term *atomos* in the fifth century BCE. Its meaning is 'uncuttable'. The core idea being that if we keep cutting something in half, and then half of one half, and then half of this again, and so on... we get something that is 'undividable' and the constituent of everything, and the variety of the world comes from the numbers of this common constituent involved.

It is again useful to illustrate this with a mental exercise. Think of halving an apple again and again—maybe to make things easier look at a human hair on your hand, look at the diameter, and imagine getting theoretically tiny scissors and cutting this in half again and again. Do we reach a point where you can't cut again, and, if so, how? What might also surprise you, or perhaps not, is that the journey to the small is as long as the journey to the big.

Using some everyday analogies, as always, let's look at the apple again, and revisit the journey to the vastness of the whole world. A football dwarfs the apple, certainly a football stadium makes it pretty small, a city, a country, an ocean, the world. The apple in proportion to the whole massive earth is the same as the apple to an atom. Perhaps that one doesn't do it for you — let's go back to the human hair again. In fact, if you have some to spare pull one from your head, have a good look at it, and try to imagine putting ten or fifty of anything across the width of it. Impossible to even see ten in your view isn't it? There are different atoms, and indeed hairs. The estimations are in the region of 200,000 to one million atoms across the hair. Again, mind-blowing I think you will agree.

The atom itself, though, is massive in the world of the small. As most everyone will know from just rudimentary science schooling the atom has a nucleus, which may be made of just one proton or a number of protons and neutrons bound together; and also, of orbiting electrons. Again though, like our planetary models, it would not be practicable for any model shown in a school or elsewhere to be anything like the scale of reality. A hydrogen atom will have one proton and one electron. Start from the size of the full stop at the end of this sentence. In this you could place 50,000 million protons. Getting that sort of dizzying feeling again? However, the proton is truly massive when compared to the electron, a mass just $1/1836$ of that of the proton.

Another pretty startling observation is the distance between them. If the full stop was the massive proton and the electron many, many magnitudes too small to be seen (depending on your eyesight, two or three times smaller than the full stop would be invisible), on these scales, where on the page would the electron be located? You guessed it, it's not on the page, it's 50 metres away. When we have a picture of this in our minds, this 'full stop' being orbited by this invisibly small particle half a football field away, what really strikes us is that matter really is not solid at all; effectively the atoms that make up absolutely everything you are familiar with are for the vast, vast majority just empty space.

This is not a thought that is too comforting if your feet are on something supposedly solid while hovering at a height; but, of course, rock to us is rock and it is plenty solid enough — even if it is practically all made of empty space. We do not take in this information and shift our behaviours, becoming afraid of a hyperactive space-filled floor. Instead we discover but react to the world as we are designed to. The thing is, of course, that we are made of these same atoms, and are a part of this particular dance. There are other incidents where having discovered something we do not apply the findings to everyday life in contravention of our experiences; two of these were generated by Einstein's 1905 *annus mirabilis*.

Einstein

We are getting right into the excitement of the world as we know it now. I am minded to say that reading around this stuff would be a good idea if you are new to these concepts, certainly Walter Isaacson's *Einstein* will combine a very good human tale with an accessible guide through the science. There are many others also. From a caricature-like idea I had of him a decade ago, Albert Einstein would now have to top any list of all time heroes for me, and it is interesting to see how many great individuals themselves in his field looked up and sought approval from this gentle intellectual giant.

Einstein introduced the idea of a quantum of light in 1905, as well as the special theory of relativity. It could be argued, due to the current state of quantum understanding or lack thereof, that in, arguably, the greatest piece of thinking a human has ever achieved he ultimately pointed the way to the weaknesses or fallibility of human thought.

Isaac Newton, who at this point was universally seen as a short second to God as an authority on the world, had explained how the world and its parts react to each other and gave us truths that will always remain, viewed in the context that they were hypothesised. Newton himself knew there was a gap in his monumental achievements and work towards understanding the universe. Central to Newton's work were ideas that, in totality, energy was conserved, so while it might be used up/spent/burnt somewhere it went somewhere else; the fire's energy transferred as heat in the room, as the kinetic energy that moves the resistant energy of the levers on a coal train, etc. Also of universal importance was the concept that objects attracted each other, and that the bigger the mass of one thing the more attraction it had — the attraction was gravity, and the motions of the planets and the famous apple all fell into place here. It might be worth considering that, in fact, when the apple falls, the Earth and the apple move towards each other according to their mass. The apple though, as we have already said, is a hell of a lot smaller and so does effectively all the moving, but if Earth was to get too close to something massive it would be the one 'falling'.

Newton's universe also had the incredibly user-friendly concepts of absolute time and space. Essentially this means that when I move, or a planet moves, we know that we move because something is stationery against which our motion can be perceived.

Anyway, back to Newton's conceded weakness; what happens if the moon that dances through gravity with the Earth all the time were to cease to exist or collapse in on itself like a black hole or something, surely then we must feel the effects of this instantaneously as we are 'connected' through Newton's laws. Connected by what though?

The idea at the time was that there was a background to everything, and that it was everywhere, and it was called the ether. The ether, then, was essentially a field in which everything else played itself out, if you will. Funnily enough, the concept of a universal field is at the moment being tested again, but this is one example of the sort of problem that faced Newtonian laws; what moved so fast as to be instantaneous, what allows the moon to disappear and an instantaneous gravitational effect be felt on Earth, what would carry the 'information' to Earth that this disappearance had, in fact, taken place? Well you might think, 'I could look up and see it disappear', even if it took a while to 'feel' it. This begs the question—how do you see it, then? The light patterns of the event, in fact, must travel to you at the speed of light. So the moon disappearing—as you see it—has actually occurred 1.5 seconds before you see it.

Fair enough to all of this I guess. However, are you happy that the moon exploded just once, or did it explode at, say, exactly 11.00 and again at 1.5 seconds past 11.00? Well I guess that depends on whose watch you are looking at.

We will come back to this in a little bit, but I want you now to jump into an elevator, a really quickly ascending elevator, the elevator is rushing up the building. You try to bounce a tennis ball off the floor; because the floor is rushing upwards the ball takes no time at all to hit the floor but pretty much stays there with very little bounce because of that same upwards thrust. Is that what happens? Of course not—the ball bounces just fine. This is because you and the ball are operating within the space of the elevator, and it is irrelevant to you, the ball, and the floor that the lift is going upwards relative to the building.

When we sit at a train station and feel that we are moving, only to discover that it is the train across the tracks that is moving really, how do we know that? Because you are not moving relative to other benchmarks (the station platform, etc.). So we can begin to see, perhaps, that motion is a relative term. In fact, as you sit and read this you are on a much larger moving object—the Earth. So if someone asked you how fast the Earth is moving, you might well answer that if I stood on the Sun and knew that the Earth takes one year to complete an orbit of 93 million miles then we can calculate that the Earth is moving at more than 10,000 miles per hour. However, we are also spinning on our own axis at 1800 miles per hour, and if you happened to be sitting towards the centre of the Milky Way you would see an Earth spinning around you at a much greater speed... And another pretty incredible event happened in 1929 when Hubble discovered that galaxies (all of them) were moving away from each other at immense speeds. We won't even get into what that said about our past just now! We can make one observation, though; if everything is moving away from everything else, then the space between things is getting bigger, or space itself is getting bigger. Anyway, for now I just mean to illustrate that it is exceedingly difficult to give a single response to the question of how fast the Earth is moving.

What we hope to illustrate with all of the above, in our mind's eye, is that something cannot be said to be moving, it is only moving relative to something else. Bear in mind that Einstein was operating with the 'knowledge' that nothing can travel faster than the speed of light. This is because a photon does not have any mass, and the greater the mass the more an object slows down its journey through space, or can be affected by gravity. So Einstein had a wonderful set of thought experiments, including one where you are riding on a beam of light — please do read as much on all of this as you can and as many times as you can. In any event, let's look at one of his very famous thought experiments.

Now you are back standing on the platform of that train station. There is a train going by the station at an incredibly fast speed, and your best friend is sitting in the middle of the train. In a rather freak event the train is struck by lightning simultaneously at the front of the train and the back of the train. Well, from the point of view of your friend in the middle of the train the two strikes were simultaneous. This is a particularly long train, though, and it is going particularly fast. As the image shoots out from the lightning strike at the back of the car, the back of the car is moving towards you and so is shortening the journey that image has to take. However, the event at the front of the long train is moving away from you and so has a further journey to travel. Both events, then, will seem to occur to you at different times. Who's right? Did the lightning strike simultaneously as your friend experienced on the train, or rather as you saw it? Bear in mind that the light that brings the event to your eyes has a constant speed. Who's right? Both of you. There is no such thing as 'simultaneous', there is only simultaneity dependent on the observer's frame of reference. If this is your first time encountering these ideas, might I advise going for a long walk to let that settle in for a while, we will be here when you return. Losing simultaneity can be difficult!

What about the flash of a star going supernovae in the sky? Did that just happen as you saw it? Well, the vastness of space that we discussed earlier and the notion of light years should give us a clue. From your frame of reference the star went supernova just now, but from the star's this may have occurred before the pyramids were built. In fact, because a distant star's light can take aeons to travel to Earth, when you look up at the night sky you are looking directly at the past. There are stars in your sky that in their own space ceased to be millions of years ago!

Just another small point of interest — if there is no simultaneity, how then does a watch or a clock work? The answer is that they don't, at least not absolutely accurately — if they did then your watch would be different after a train journey than the watch of the person you departed from. However, with the top speeds we can go, it would take many lifetimes for the differences to be perceptible. The everyday world is not really affected by these effects. Perhaps the best way to enter the strange universe that we inhabit is to consider that, if you are parked in your car and someone drives towards you at twenty miles per hour and you proceed to start your engine and drive towards that car, you are going to increase the impact of a crash as you have increased the speed at which that car is moving relative to you.

One of the two craziest concepts we have to encounter in our view of the world is that if you move towards the light from the Sun, it does not speed up. What happens? Well, if you move quickly enough to have any real impact on something that measures in the realm of 300,000 miles per second... well... time slows down! Now, just about everyone struggles massively with this concept. We are pretty dependent on a constant notion of time.

We also know now that motion is just another manifestation of gravity. Adding to his earlier special theory of relativity comes Einstein's 'happiest ever thought' as he looks out of a window at some men working on a roof. He imagines one of them falling. In Newton's world gravitational pull would be the reason this man moves towards the ground. It struck Einstein that the man would be—during the fall—weightless. Einstein now imagines that in an otherwise empty universe the man would no longer be falling (because there is nothing to fall in relation to). So Einstein realises that the Earth's mass is curving space and so space is pushing the man towards the Earth. Mass moving curves space-time around it, or behaves as gravity. And so gravity can be equated to motion. This curved view of space-time around the Earth, or the Earth creating the curvature by its mass, can be perhaps more easily related to when we think of the Earth orbiting the Sun. The old view is there is a connection between the Earth and the Sun, in fact the Earth is travelling in a straight line, but the space around the sun is curved by its mass, so you might picture in your mind's eye a sort of spherical velodrome for the Earth's motion 'around' the Sun.

When it was famously joked that only three people understood relativity (back when it was an extremely popular scientific breakthrough), it was acceptance more than understanding the detail that was the problem. This is completely counter-intuitive, but a big step forward is to appreciate that not everything that makes sense from your experience makes sense scientifically. I am sure, if this is new, you are still having major problems with accepting this—but here is a good bit of reinforcement that will assure you as you encounter relativity again and again. You know how your satnav works—beams are sent between your car and satellites that are orbiting the Earth, and by bouncing off a few together your location is read. Well, this effect that we're discussing, that time moves slower on Earth than it does on the satellite, is accounted for when the calculations are being programmed. It's just a number of microseconds, but if it's not included then the accuracy when brought back to the location system would be out by approximately 10 km a day. If you are trying to get somewhere this is certainly a significant enough error to set you very much in the wrong direction. So there we have day to day evidence of the universal weirdness that comes from our twentieth-century understanding of the relativity of existence, and the lack of absolute time and space.

However, it might be best for us to return to the world of the small, where it gets a whole lot weirder. So weird, in fact, that Einstein, again the father, could not abide the oddity of his own creation. If you've never heard of the double slit experiment before you're in for some fun!

What the ... is Quantum Mechanics?

The fun stems from the difficulty of light behaving like a wave (picture ripples spreading out uniformly with peaks and troughs when you drop a pebble in a lake) and also like a particle. From the wave-like perspective bear in mind the idea of another pebble being dropped at another point in the lake — as the ripples move out from there the two sets of ripples eventually reach each other, and as the peaks and troughs interact at some points they reinforce each other and at others they cancel each other out.

A pretty massive contradiction occurred at the beginning of the twentieth century when it was observed that light also behaved like a particle; so sometimes particles bounced off of a photon of light like two billiard balls hitting off each other. This wave–particle duality was introduced in many ways by Einstein concluding electromagnetic waves could only come in packages he called quantum.

Wave–particle duality may seem a harmless enough concept — but, again, if you haven't encountered this before, fasten your seatbelt, open your mind, and (try to) enjoy.

We have two sheets one after the other with two slits on the first. We shine a beam of light on the first sheet. As a wave the light spreads out from both slits and then both waves interfere on their route to the second sheet, thus forming a pattern of dark and bright bands on this second sheet. All of this is consistent with light behaving like and being a wave. As mentioned above, though, we know we can bounce electrons off of the particle of light. Now, if we were to slow down the intensity of the light — all the way down until there was just one photon going through the first sheet at a time — then it becomes clear that the photon is interfering with itself, as an interference pattern again becomes evident. This means that, in some way, the single photon goes through both slits, causing two waves and the resulting interference!

It is advised that you go and read as much as you can about this, have some time contradicting it and questioning the credibility of the test and the testers and find your way towards a resigned acceptance that the basic underpinning of the world is an extremely alien and weird place. Reality is a fuzzy, blurred, random, and predictable-only-on-probability kinda place. Beyond certain tiny sizes and incredible speeds this randomness disappears, not because the fundamentals change, but because of the sheer numbers involved in the probabilities.

You may well feel an urge to think, 'ah, I'm not reading any more of this — it's pointless and wrong, even if it is agreed by all scientists for almost a century now'. This is your internal representation getting a bit of a thumping. If you are feeling this way then skip the next couple of sentences. When you slow the photons down so much that only one photon goes through the first sheet at a time, if you observe the photon it will let you know which slit it goes through. If you don't directly observe it it changes its behaviour and seems to go through both!

Entropy

Another major theoretical breakthrough of the early twentieth century was presented by Ludwig Boltzmann just prior to his tragic suicide in 1906. Boltzmann suffered from bipolar disorder and had difficulties keeping the order on his own world interpretation. Again, this type of tragic disease and possible future understandings will be among topics that will be discussed in some detail later, when looking at connectomics and other rapidly growing fields within the study of the mind. His most significant breakthrough might be described as emanating from considerations of disorder.

Boltzmann's and others' insight was that, when we think of it, it is order that is highly unusual. If you drop a glass on the floor it breaks into a whole big mess. Now, it could be that after breaking and smashing into many pieces which bounce off of each other repeatedly (all of these motions being contributed to by the temperature, the substance of the floor, the type of glass, height and angle of the fall, etc.) the glass could come to rest as a smiley face, or to spell your name, or to reconstruct itself as it was prior to the fall. Ask yourself which of these you would see as least likely to occur, and second most likely, and most likely to occur. So for the sake of argument let's assume you selected most to least likely in the following order: smiley face, then your name, and then rearrange into prior form. Why did you select that order though? The answer, of course, is that each of these requires more order, more exactness, than the previous one; and so it seems that order is unlikely.

In fact, the probability is much higher that it just lies on the floor all messed up than any of the above scenarios. Why? Because there are billions of ways that the glass could fragment and come to rest in a mess; there are many fewer ways this could occur so that it could be seen to represent a smiley face, and only one where it could rearrange itself to its previous state. So this is all down just simply to the numbers. Roll a dice and you have a chance of getting your selected number, choose the same number as a randomly selected number from one to ten trillion, and we are pretty much safe in saying that you won't get it right. This natural statistical preference for disorder is called entropy.

The general tendency in the universe is towards entropy (simply because it is more probable) and is where an expanding, cooling universe seems to be heading—total isolated disorder. Where Earth, ants, and humans sit with this tendency towards entropy is difficult to explain but, in general, while we take food sources and process them for growth and development and sustenance, and interact in a variety of different ways with our environment, including emitting heat from our bodies throughout, believe it or not we, like everything else, add to entropy over the course of our lifetimes. But how incredibly important the essence of the order of which we are made, DNA, is — the building block of another massive revolution of understanding — the understanding of our own biology.

Chapter Three
What is Stuff Made of?

The Same, Only Different!

How do you differ from a monkey or a block of cheese, or a rock or a star, or, indeed, a rock star? Well, a marvellous thing now is that we know of what we are made—and, while we haven't completely figured out the instructions, we now also know roughly speaking how we are made. If there is something non-physical or spiritual, we can leave this aside for the moment to at least see that for the most part it is a question of quantities and mixes of the same central elements, of which there are 109 stable forms (however, there are more being found, with more and more ridiculously short life expectancies, in experiments around the world). It is important to note that many elements are extremely rare; some are very rare on Earth, and others are rare right across the universe. Hydrogen, helium, oxygen, and carbon are abundant throughout, and carbon particularly is essential to what we call 'life'.

Bare Necessities

We mentioned that everything we know in the world around us is made of 109 different types of atoms, and we highlighted that atoms were made of electrons and nuclei consisting of protons and neutrons. In fact, particle physics takes us much further down the ladder of size or mass beyond that—essentially, though, what particle physicists are telling us is that there are incredibly tiny particles that form together and come apart in certain extreme circumstances (many of which are being recreated in locations like Cern). For a long time it was felt that there was a backdrop to these particles known as fields, and to an extent the goalposts are being redefined and the recently confirmed Higgs particle corroborates this. I don't think we need to look at string theory just now—but essentially in this broad model these particles or points are the tips of vibrating strings, the strings themselves being curled up in dimensions that may remain impossible for us to see.

Anyway, string theory or fields aside we can still suggest that, be they the fermions or bosons, these subsets of particles themselves are identical to each other—indeed, so are electrons. There is no difference between this electron and that electron. From the mixtures of these elementary particles and their varying quantities comes existence. At one point in my own mental journey, as I was wandering down the various philosophical ramifications of all of this, it struck me, at least, that the differences that emerged were pivotal.

If everything were one thing, I thought, there would be no reference point. How could we describe Penelope Cruz if there was no *other*—more to the point how could the differences that make up her, 200 different cell types and organs, etc., be if there were no fundamental differences or variety available to us. Indeed, what about a rock and the *empty* space beside it—how could both exist unless the constituent parts differed? So if everything had the same underlying source, in short, there would be nothing. The difference allows for comparison, identification, and interaction. I further battled with the concept of continuous 'halving'—perhaps we can theorise our way down to ever smaller particles (though we seem to be at mass-less energies now). But let's say we don't definitely know how much more room there is at the bottom. What we must continue to find is some trace of difference or, as I say, there would just be sameness, and no way we could arrive at our world.

Chapter Three: What is Stuff Made of?

What's the Matter?

Sometimes suddenly something just jumps up and hits you between the eyes. I read in one book about this guy saying that when he was a child he heard that $1 + 1 = 2$, and then at a later stage he got it, he really got it. To be honest when reading it I didn't understand what he was on about myself, but then something very similar occurred to me. Just when I was pondering the above point about the centrality of difference as we peer down through the smaller parts of what stuff is made of — and working on the assumption that we can't go all the way to nothing or even just one thing — then what underpins this must be the holding of the instruction that gives the difference, and so, ultimately, eventually, must lie *information*. ***All matter is information.***

In the end, I thought to myself, all things do not merge into one — the information, the 'a little of this plus some of that and a tiny portion of this… that interacts with the pull of this and the push of that… and so on' — this information, or recipe, is what underlies everything. You may fret at questions like 'where did the information come from?' now, but that is not the point here. It can be the result of a cosmic role of the symmetry breaking big bang dice, or you can feel God's hand or another 'intelligent designer' if you wish — what is important is the appreciation of why we can say that everything is information.

This is a phrase I had heard previously and, like the maths enthusiast above, at that point I felt like I got it, I really got it; and I got its significance.

Out of a sea of potential nothingness comes the information that encodes the universe, and that is computed (by computed we need to appreciate we mean 'occurs' here, basically) by laws of entropy, probability, gravity, quantum mechanics, and the arrow of time. Of course, what also jumped into my mind at this stage were the informational, traditionally-encoded manner of DNA and the history of computers, and how these two could relate to the Eureka moment above. This permeates the rest of the book.

Chapter Four
All About Us

What is Life? (BIOLOGY – HEREDITY – ORDER)

What is life? Edwin Schrödinger wrote a wonderful book with the title *What is Life* in 1946. Schrödinger was already a famed physicist with his place in the history of the origins of quantum theory firmly established, via the central Schrödinger's equation and, in some ways even more famously, his fun thought experiment dubbed Schrödinger's Cat, in which, in general support of Einstein's contemporary challenge to the prevailing quantum interpretation of the world, he highlighted an experiment that might leave a cat both dead and alive at the same time—if the logic of quantum mechanics were to be followed! It may well surprise you that in the end people sided against Schrödinger and Einstein in this whole debate—the Solvay conference's wonderful exchanges between Niels Bohr and supporters, and Einstein himself, on the nature of reality really might make particularly good reading for you if you haven't found them already.

Schrödinger took his 'essential' physicist's knowledge and understanding and, not unlike others, turned his mind to biology. In this fascinating, if a little detailed, book Schrödinger, many believe, foretold the role of DNA from a philosophical perspective seven years prior to its discovery, and focused on the structure required for life and the means to retain and replicate that structure. Now it is important to consider entropy again—didn't Boltzmann explain that the logic of probabilities pushes most things towards disorder, because there are far more ways to be disordered than ordered?

The answer is yes; but what we must bear in mind is the scale in space and time, or space-time, of the universe, the observable universe and otherwise and the singular or theorised infinite numbers of universes.

Now, in that context, how you beat probabilities is to keep trying. So if you keep breaking glass eventually the numbers will work in your favour and something unusual will happen. The something unusual here, in the context of chemical patterns and the bedrock of life evolution, is the ability/propensity for a particular pattern to dominate competing completely random patterns and become a repetitive form. For example, if you let a completely random set of balls interact with each other, while it is incredibly unlikely that they will hop and bounce and rest spelling the word pool by their positions, it is also an unlikely proposition that they will split in a uniform spherical and equally separated manner. Therefore clustering that might in the case of gravity, for example, force other items into the new found unequal patterns of distribution can all combine to give us what appears to be a shape emerging from no instructions, but is merely the result of probabilities in all its forms. In some cases complexity becomes inherent in the form and it increases from there by self-propagation.

From gravity and grouping and patterns over iterations we arrive at a universe that has been explained to a great extent now by scientific endeavours through the eras; from atoms to stars (or stars to atoms) to planets, to the embryonic biosphere, and to this incredible and absolutely essential story (for us) of evolution, that stopped and started a bit but which contains so many answers that many still insist on seeking elsewhere. Patterns of information are central but many argue that there are known unknowables in this.

Can we recreate precisely and totally comprehend even a storm cloud, they may ask? Theoretically yes, but practically no. If we had total information of all components and an endless amount of computational power, then yes we could appreciate all parts. The adherence to absolute fidelity will be raised again in Book 3 but I would ask us to consider now, for example, would we need to appreciate the bonding of oxygen and hydrogen in order to grasp the fundamentals of a water molecule? Yes. But if I see a litre of water spill and I am thirsty, do I need a subatomic level fidelity in recreating this litre of water so I can drink again? Not at all—I just need to run the tap a little.

While this is an example of extreme redundancy, is there any relevance to our understanding the human brain? Is it 100 billion neurons that all need fidelity, or 100 trillion synaptic connections between those neurons, or might we have gotten a little carried away with our own sophistication? I know that I for one can't comprehend the number 100 trillion, so while it does sort of make me think, well—you know, I am no worm or other simple creature, but (and it's a pretty massive *but* for me) am I in any way complex enough to require 100 trillion of anything? Welcome to the brain, the seat of consciousness, the key to the subjectivity trap.

What Are We? (BIOLOGY)

Before we explore the brain, let's try to appreciate the constitution of our bodies, which can malfunction in so many ways that we must appreciate we are dealing with a pretty awesomely complex entity. However, if we can park the creationist theories let's explore the core concepts of evolution.

It is very impressive how birds fly, how ants cooperate, how giraffes grow and move, but it is something that is a *result*, it is something that was not designed or pre-planned. It is not perfect; it is, as with all things that come through the filter of 'fitness', useful. It will also (for the most part) advance in as simple a manner as possible. Layer upon layer of simple units following simple rules can manifest over aeons as complexity, and this is what we find in organisms, but it should also leave the relative simplicity at the underlying levels when these are appreciated. For most of human history little was known about these underlying biological units.

DNA

You, or I, begin as a zygote, the primary cell resulting from the fusion of the sperm and the egg, the holders of the germ line.

There is a pre-programmed development from this single cell to the totality of a mature human. All cells carry the same information, or DNA. There are four different nitrogen bases: Adenine (A), Thymine (T), Cytosine (C), and Guanine (G). In the deoxyribonucleic acid, these four base code proteins come in sets of three and (as anyone who bets multiples will tell you quickly) there are potentially twenty different amino acids formed from these sets of three. Think about how many combination sets consisting of three numbers you can make from the numbers 1, 2, 3, and 4. 'So what?', you might well ask. Well, these amino acids then fold into one another in a variety of shapes that create the variety of cells that we may find in our beings, from our skins and our skeletons, our muscles, our bodily systems, and our organs—everything that is us is dependent on how the initial bases are laid down.

The two strands of DNA that are matched in the now very famous spiral double helix split and each individual strand acts as a magnet (with help from DNA polymerase) to partner bases again. So there are now two individual strands and they are matched by a very simple process—where, as explained, T is matched by A and A matched with T, and C with G and G with C.

If this is how the two strands are matched up with new partners to now make two new cells, you might well ask where the new material comes from. Well, it is stored in the cell and written across by complex cellular mechanisms, but, importantly, where it originates is that it is ingested. So the phrase 'you are what you eat' is in fact correct. The digestive and metabolic system first breaks down what is ingested through a series of steps to leave the materials required—so try not to imagine your DNA as being formed in part from a series of fatty bacon!

There are 6 billion pairs in a DNA strand so this is by no means a simple matter. But at least each cell uses the same code. And as different as a cell in the eye and the liver and the brain may seem, it is nice to know that they are each formed from the same 'text' or code. The differences come from different parts of the code being active or 'switched on'.

The proteins fold in accordance with the active elements of the code and the base cells take on particular functions, such as to form skin, or heart muscle, or teeth, or neurons. Collections of these cells (lots of them) take on the shape of our main organs; lungs, kidneys, stomach, brain, heart, gallbladder, intestines, spleen, liver, skin; and the thirteen bodily systems, such as circulatory, lymphatic, endocrine, respiratory, reproductive, etc. The musculoskeletal system with its tendons, ligaments, cartilage, and bones provides the 'scaffolding', and in many ways illustrates the commonality I refer to as in these cases we can see that the essential fibrous commodity is shared — it is, for the most part, the density that differentiates between muscles and bones, and so on.

There are many subsystems in this human entity, and gargantuan volumes of crossover from system to system and organ to organ, and so results the encyclopaedia of medical knowledge and skills we see in modern medicine. We also see that there are, again, scales of impacts. So, with a major trauma, as with a fall from a major height, the integrity of the whole system will be destroyed. We may also see that, if there is tissue damage to a pivotal point in the musculoskeletal structure, this can harm the body in a variety of ways and leave someone suffering with 'a bad back' for years (for example), as the body accommodates discomforts and so places pressures on other parts of the structure, and so on, until eventually increased stresses and absorption of medication for pain treatment may lead to a point when the individual suffers from a whole range of other difficulties.

However, in only the beginning of mankind's second decade of gene sequencing, and our sixth with appreciation of the centrality of DNA, we see where stem cell treatment is heading — where skin cells now can be coaxed into reforming into specialists such as nerve cells — and, from being injected into the generally affected area (in a largely haphazard manner), these new cells can take the place of the damaged cells. This is all in its infancy and is clearly very hit-and-miss in its application at present, but these minor successes are very strong proof of concepts.

In the fullness of time the appreciation of DNA may be seen as being as important as Copernicus showing that the Sun did not revolve around the Earth in terms of a step towards our gradual self-demystification. Most still see the human entity as incomprehensible, mysterious, other-worldly, and so on. In many cases they will state that if it was possible to understand the human being we would have done so by now.

Well, no! Certainly not. We didn't have more than what the naked eye could discern until the sixteenth century with the early microscope. This line of conversation could be extended. Remember that we are gaining knowledge here of a particular thing—that of human anatomy and physiology; and so as knowledge accumulates unknowns decline… But more than this—as well as the awesome processing power of computers of recent decades, we have also seen incredible leaps forward in miniaturisation and imaging techniques. These have altered the resolution at which we can see the human system.

The more we see, the more function reveals itself. Far from our bodies being the vessel for our infinitely important mind, it seems that the human entity as a whole appears to treat both thoughts and digestion as processes requiring energy. It contains 206 bones in its skeletal system and, when combined with its more elastic and less dense fibrous muscular system, the robust structure of the individual stands. It is interesting also for us to see that, as mentioned before, bone is really just muscle condensed into a more solid form.

Within that structure a whole range of organs and systems interact; and, undoubtedly, as with any diverse collaborative network, there is almost a limitless pool of potential faults therein. However, when one looks at what the majority of the system actually does, we see something that digests looking for energy sources to operate control over all the muscles in its own system and also cellular rejuvenation.

The waste from these processes are, at least in a properly functioning system, then dispelled in a very non-ceremonious manner. Huge amounts of the human machinery are invested in maintaining its well-being. There is a massive amount of information used to evaluate that well-being and many, many 'evolutionary' decisions seemed to have followed the lowest energy cost route throughout our developmental history. We really do seem like energy users, and one can look at all aspects of us as a means to find that energy, such as the giraffe's long neck—but more of that later.

But how do we address the brain, in all of its wondrous complexity? And beyond that how can we possibly explain the inexplicable, the centre of high-end thinking, the location of creativity, the 'I', or consciousness?

The Brain

CBT and Consciousness

What in truth is going on inside the skull in the relative mush pie that looks like, well, like brain, I guess? It seems to be a soft, soupy-like collection of not a whole pile of significance. In fact, using autopsies it is not that surprising that medical opinion held for aeons that the heart was the centre of thought, emotions, and so on—and still remains as such in the general lexicon. The fact that emotion in some ways resides in our heart (e.g. 'my heart broke') results from a connection between the heart and the brain that is actually very real. So, when our mind is encountering anxiety, fear, frustration, etc., our heart's rhythm actually changes and becomes erratic; following from this, hormones are released, blood vessels constrict, and so blood pressure rises. These are all parts of the evolved alarm systems that kick into activity as a model of the ancient fight-or-flight mechanism. Conversely, if you feel in love, at ease, contented, your heart's rhythm is more harmonious, it becomes more efficient, serotonin and an array of other chemicals which are deemed good by the brain and are made manifest by what we call 'happiness' follow on.

Chapter Four: All About Us

As we begin to glimpse the physicality of our mentality, we begin to see that much of what seemed to come from the mysterious throughout the ages was actually the correct result from the wrong understanding. That is to say that meditation, peacefulness, appropriate exercise, and contentedness can all increase your well-being, though not because it empowers your soul, but rather your anatomical systems! There are many more examples of physical supports given to seemingly mental activities (I will continue to use physical/mental as separate terms just for a little while more) — to name just a few; the rhythm and rote nature of a Gregorian chant, the consciousness numbing feeling from a repetitive prayer, as well as the predictable nature of the angelus bells, and the quiet and incense-filled atmosphere of a church or other prayer building. In all these cases there are proven and repeated environmental impacts on our mental states.

In fact, the field known as Cognitive Behavioural Therapy (CBT) is a fascinating development within the 'lower resolution' attempts at comprehending thought. Freud and Jung and the vast fields of psychoanalysis should not be dismissed here — in these cases thought was being understood and affected by thinking (or, to put it another way, thought was being understood at the level of the conscious 'I'). There is a level of functionality below this, however. The lower functional level impact of CBT does in many ways point to the illusion of conscious autonomy, and the simplicity, in fact, of the means of addressing some mental illness, particularly the likes of depression, phobias, and anxiety.

I might mention here to any reader that has had some form of therapy, you are much more likely to be able to get through the concept of the 'Subjectivity Trap' precisely because the confidence in your own 'voice' has been at least shaken a little already.

For example, we can look at the case of anxiety. Returning to your evolutionary ancestor, he/she lived in very dangerous circumstances (like many predatory environments that we view on safari trips these days!) and he/she had a very centrally hard-wired reaction to danger that we now call 'fight-or-flight'.

This reaction involved a coursing of adrenaline, a heightened awareness of bodily sensation and environment, a desire to move, and a desire to empty one's bowels in order to increase the speed of that movement—the origin of the phrase 'I was shitting myself', I hear you say. Yes, that would seem to be the case.

Modern humans all still have this central mechanism, and it is fairly deep and strong in terms of the control it can exert over our actions, in an instant of activation. Think about how quickly you moved the last time you felt something was going to hit you, and how quickly your interior monologue turned off. Also, recall the last time you were really sick, remember the moment when it occurred to the entirety of you, that there is something wrong here, wow, what is this, why do I feel so bad, and a rush of adrenaline and panic coursed through you, and your mind sought about rapidly seeking to locate the cause of the problem (akin to seeing blood on your shirt, and desperately trying to locate its origin).

These are all perfectly useful reactions. They are alarms of the human system, if you like. It would make perfect sense that any system charged with defending its own self would have evolved these warning systems. The issue with people suffering from phobias and general anxiety is that the alarm is going off unnecessarily, or the alarm system is broken. With the condition known as Generalised Anxiety Disorder, over a period of time the frequency of the alarm sounding can be increased until it basically goes from one cycle to another without any break. This individual then is imprisoned in a neurochemical malfunction that has them feeling the sensations of what we discussed at the horrible moment of first illness awareness many times a day.

A loop of inter-neural connectivity, with increased plasticity (we will cover all of this soon), builds up and so each occurrence enhances the likelihood of it happening again, and so on. The conscious 'I' runs around the place looking for the source of the adrenaline rush, palpitations, etc. Unsurprisingly when other symptoms can include tightness in the chest area and breathlessness, victims often feel that they are having a heart attack, which can, undoubtedly, cause its own round of panic.

How does one break the cycle? Well, when it is deeply ingrained and the conscious 'I' really is only a spectator to these loops, then medication may dampen all activity, including this broken alarm, and allow time/room to further develop coping mechanisms. About here we can meet CBT. The body and mind try together, under expert tutelage, to break the loop. And, while it is certainly not instantaneous and requires learning and reinforcing in itself, a 'better' or more useful interpretation of reality can return.

Meeting Consciousness

Why so much detail about something so specific as CBT to treat anxiety or a phobia, or depression for that matter? Well, read back over what we have just gone through. There is a pretty key point here in our attempt to escape from the subjectivity trap. Your thoughts and your physical sensations are working in tandem. One influences and is influenced by the other… Because they are the same thing.

There is no mind–body duality — this is a construct of what we deem to be 'the mind'. It is all biology!

Book Two

Chapter Five
What is the Subjectivity Trap?

In this section we will look at answering this question from a range of different perspectives.

This chapter may jump around quite a bit—feel free to do likewise.

Purkinje Fibres

In the heart muscle there are specialised fibres called Purkinje fibres. They are actually modified muscle tissues that can carry electrical signals. The fibres are intricately linked to the heart muscle. They allow coordinated contraction of the heart muscles to support the vital blood flow through the body. The signals for this contraction come from the autonomic system, i.e. what we feel is something outside of our conscious decision making control—try telling your heart to stop beating!

There are neurons all around the body, but those in the brain are part of a vast mesh of connections while those outside the brain only connect to a few others. The vast majority connect to other neurons, including the majority located in the fulcrum of the brain. For those that are located at the nerve's meeting point with a muscle, their conclusion is a motor end plate. This operates in a similar manner to other neuron-to-neuron connections, that is to say chemicals cover what is called the synaptic cleft (a little gap between the neurons—or, particularly, the point that *joins* the synapse to the next neuron) that tells the receiving cell to set off and have the signal continue on to the cells that it happens to be connected to — cognisant that, for one to fire, a certain amount of the many that link to it must fire. However, at the motor end the chemical 'connection' causes reactions that lead to muscle contraction or relaxation. This is all very clearly physical, simple interactions common to vast tracts of the animal kingdom.

In the brain itself, of course, there are large tracts of sensory information entering the system of neurons 'firing', and so setting off other surrounding neurons all the time; this includes the nerve endings on the skin that allow us to 'sense' touch, in the nasal passage for smell, and so on. Remember when you walk out into the rain and your skin senses the water drops? But we are talking about literally millions of nerve endings here, all encoding what we (the subjective version) would deem to be one message — 'it's raining'. This distilling or compressing of information by the subjective 'I' happens all the time. In fact, the numbers involved in our comprehension, one might say, are beyond our comprehension! So it seems that there is some serious computation occurring in the brain and nervous system, when all of this sensory information is processed there are corresponding pre-programmed motor reactions. AND THAT'S IT! THAT'S HOW THE BRAIN WORKS.

'Wait a minute!', I here you shout. What about thinking? Hopefully you can see over the coming sections that it is just a question of connections (in fact there are many academics now using the term connectomics for new interpretations of the brain). We know that the neurons operate, at one level, in a relatively simple arithmetic known as Bayesian logic. This means that when faced with a number of options the computation basically tries to discount its way to something that closely matches the inputs — looking for the *'less wrong'* option. When there are many raindrops the fact that it is rain is reinforced again and again and the notion of something falling that is heavier or colder or more dangerous is discounted. How do we know it's 'rain'? Well, we don't initially. This is a word that language acquisition allows us to associate with the sensation—our visual sensory system is also probably scanning the sky for a whole series of possibilities and also arriving at the same conclusion. Of course, we may go into the rain because our child is outside and in danger, and there are socially learned and instinctively innate protective templates that are kicking in then also.

Now we know how the Purkinje fibres work, we know how neurons work, we know how Bayesian logic works, we know that we infer all the time, make predictions (we know, for example, that our retina only receives a fraction of the visual picture we constantly infer by prediction—or we see with our eyes only part of what we *see*, we fill in the majority of it in supporting ganglions and specialist processing areas of the brain—basically guessing what remains of the outline of a picture our retina receives); and we know how neurons communicate, we know how nerve endings send signals throughout the body and brain, we know how signals from the brain control muscle contraction, we know this governs movement, dexterous and clumsy activities, nail and hair growth, breathing and indeed the monitoring and management of systems such as the circulatory and respiratory systems. You would be amazed how much of something like vision neuroscience now understands—even computational neuroscience is beginning to frame models of how the brain generally adds information and computes-interprets-reacts—but it is based on facts, simple facts like a particular neuron firing when an object in a certain visual field is at a particular angle or closer to that angle than not; if the item is at a different angle, this neuron simply doesn't fire. We know that this firing is accumulated within other frames where it is treated like a node and so a 'picture' is 'built up' in the vision field. We know much about smell and taste and hearing too, by the way, and a whole range of other functions. In terms of the difference between understanding and accepting I think retinal and cochlear implants speak of a world that would have seemed ridiculous to any previous generation in this frontier.

We pretty much have everything we need, and yet we repeatedly say, we all say all the time, that we have no idea really how the brain works. Surely it is incumbent on us to think about this again, and see if we can build from the bottom up and consider: if something so well understood at so many levels remains completely mysterious, could we possibly be operating under a false fundamental premise? Is it that the subjectivity trap and the associated egocentricity have forced us to start with some of the outcomes of this neural activity and just claim it's impossible that this can result from the underlying functionality which we already understand?

I believe there really can't be another answer. To square the circle of functionality with all that we see as abstract, creative, spiritual thinking we must just take one step at a time, without making any unwarranted assumptions. Let's look at the physicality of speech. The diaphragm and vocal cords, mouth, and tongue respond very finely to motor nerve instructions; different muscles relax and contract, again through the same process. This language is pretty amazing in terms of the control; the range of sounds that can emerge through the breathing and the vocal cords is surely impressive, like many muscle control systems you may wish to admire in fellow creatures — a diving bird of prey's control of flight, for example. One step we must take is to appreciate the role of speech, and more broadly communication, in our evolution and the particular role of the 'narrator' of the communicative function, again, as, and how, it has evolved.

Starting a Conversation

The delusion of consciousness is one of the remaining grand starting point errors in perception from which our species is gradually distancing itself, and like any error of perception it can be useful at times and harmful at others. Consider that at some point in human evolutionary development one of our ancestors was less developed intellectually than a dog. In fact, at some point our ancestors were single cell organisms, and so the dog is one of the more advanced intellectual benchmarks. We were all once LUCA (the Last Universal Common Ancestor). Though, in truth, we rarely even acknowledge this — we state that we were all monkeys or evolved from monkeys (ape-like primates) (some of us can't agree to this, regardless of evidence) but when do you hear conversations around our reptilian pasts? From our lofty perspective we are, of course, adamant that there is a major difference between us and a dog, or most certainly between us and a single cell organism; for the most part, we identify that difference as consciousness.

I am aware of me, I am aware of being conscious, I am one entity, I am capable of planning into the future, I am conscious of time elapsing, I think, I think about thinking — I think, therefore I am. We see the dog as not conscious or, certainly, less conscious than us. Sure the dog has a long-term memory and will recognise you if you haven't seen him/her in a month. However, we don't see a dog checking its watch anxiously wondering where you are and what you're doing; and why it exists!

Many people believe that, although they can understand biologically the operations of the single cell, and, indeed, the physiology of the dog, and in more detail the neural control of that physiology, they are still adamant that they cannot understand the nature of consciousness, and some believe that it will never be understood. This is surely a strange standpoint.

So, if you accept that we were once as simple as creatures that we would now deem to be not conscious, let us look at this perceived unbridgeable difference between the type of creature we were and the type of creature we are, and see if we can find the magic!

When a dog (and a whole range of other creatures with similar actions) wags its tail, for example, they are using motor neuron signals carried from the brain to allow for muscle movement and other physiologically controlled uses of energy that their bodies absorb and consume for as long as they remain alive. The neural correlation of these movements is becoming better understood all of the time. The fact that this becomes a learned behaviour or a memory has been shown by the actual direct study of neurons in simpler creatures, such as when Nobel Prize winner Dr Eric Kandel observed the synaptic connection strengthening by means of protein generation between neurons in the much simpler (and easier to get ethical approval for) studies of the marine mollusc gill withdrawal reflex. As a stimulus caused the gill to withdraw, the involved neurons fired repeatedly and the connection strengthened, generating a circumstance where the more often the connection was activated, the more easily it would activate again when prompted.

So the dog wagging its tail is somewhat more complex than the gill withdrawal reflex but there are a few more neurons in a dog than in the mollusc, say 20,000 to 200,000,000. Of course, we have 100,000,000,000. What we can surely accept is that nothing magical takes place between the learned gill withdrawal reflex and the tail wagging, just complexity, and complexity that can only emerge from simplicity on top of simplicity.

What about when a dog barks? The system is the same as when its tail wags (and indeed as when the gill withdraws), but instead of just muscle control of the tail we now have utilisation of other parts of the physiology that developed through evolution. The diaphragm and the oxygen-generating lungs are employed to expel air through the vocal cords to emit a learned — as well as inherited DNA-passed attribute-based (this simply means that, when something happens by random and is useful by simple mathematics of greater numbers surviving, it is retained in DNA to a greater extent — this has useful manifestations over very long time periods and very, very many generations) — useful — sound. Following on synaptic learning and chemical-based releases such as serotonin, the dog may emit what we would call 'joy' within the particular resonance of a particular bark; or indeed fear or anger, etc. Why does a returning owner elicit joy and an intruder elicit anger? There will be a clear link found between chemicals of fight or flight and the activation of neural networks that lead to the aggressive bark, and more related to positive reinforcement related to the 'happy' bark. These emotive barks again will employ both the many-generations approach and the individual-learning approaches. Essentially, though, they can be pared back to where the dog initially receives food and desired care. An owner to one dog is an intruder to another—the system itself has evolved by innate self-preservation and also, of course, the dog's individual and special history of assisting and receiving reward lies in the hardware of the DNA-based, protein-encoded brain structure as well as the environmental dynamic element highlighted by the synaptic connection alterations as with Kandel's work.

Alright, but surely the dog is very different to us; even if we were at one stage less evolved than that dog. Surely something mysterious, magical, or impossibly complex must be taking place to explain our consciousness?

Language

Earlier humans would also have 'spoken' in a similar manner to the dog barking, and the guttural, primal sounds would have initially carried similar intentions and emotions as those we outlined for the dog. Our distant ancestors may well have jumped around and grunted when their tribe were under attack by a beast of prey. Over time, repetition, and species learning one sound would have been taken to represent a particular beast and a particular circumstance. And over the vastness of the timescales involved (for humanoid and human/ape-like creatures—perhaps 1,000,000 years) there was a major advantage generated by our species' increasingly complex means of cooperating and communicating. Working together and responding to one of the core drives, empathy (see Jeremy Rifkin's RSA youtube 'Empathic Civilisation'), humans fought and survived in groups, developed weapons and transport, medicines and language.

It is, undoubtedly, very difficult to see how these ancient guttural sounds might evolve to subtle rhetoric but, as stated, these are a vast number of lifetimes and genetic inheritances, as well as gradual complementary complexity evolutions of the brain itself—again, of course relying on the natural selection discussed previously.

We might still ask, though, is language not less than thought? Well, when we take language as a motor function (as with a dog barking) and we analyse what we know occurs in the brain with more traditional motions in the motor repertoire, we see that there are, in fact, a number of little delusions built into our overall system that also come to bear. Centrally, we are not aware of the fact that before we do something, like lift our hand, we practise it! The event takes place with a magnificent complexity and involving a vast number of neurons firing in our supplementary motor cortex prior to the motor cortex. It is a rehearsal. It occurs in the supplementary cortex and then afterwards in the motor cortex. But, crucially, this 'I' we are familiar with is not aware of the 'rehearsal'. As an aside—*there are many studies that illustrate this earlier event, perhaps most famously by Libet, and while some are disputed and critics are scrambling around to see when consciousness is alerted in different ways to try to rescue its authenticity—many studies show a delay between the origin of a movement of a finger, for example, and conscious awareness of this. This is weird right? Finger begins to move—I think I will move my finger! Well, like many other things it makes sense in the context of the useful illusory nature of executive authority—the 'I'-ness that the subjective trap has engendered. The 'I' being one mental function that claims an executive position—a total authority—it doesn't have.*

Return to our far distant relation on the African plain, that is born and living in a time before any great complexity in language but is aware (through experience/exposure) of a particular motor-driven primal sound that (having mimicked during their development) notifies others in their group that a beast is behind them. Evolutionarily this gentleman also takes great pleasure from (or serotonin-released encouraging reproduction and species survival) sexual activities, and has one particular female tribe member to whom he is particularly attracted. Unfortunately, there is an alpha male who has been getting in the way of this attraction. Our friend, at this early stage of sensory intake and motor response gaining complexity into what we now perceive as thought, is about to make this sound when a secondary drive which is to kill or get rid of the tribe's member that is denying him his sexual relations kicks in, and he does not go ahead with the motor-driven sound. Complexity has been added and he has 'thought' his way through the situation by not utilising the learned motor response. Of course, many billions of lifetimes later we follow on from civilisations that have come and gone, all the time communicating and refining the motor tool of language and its supplementary rehearsal of thought, and on into written communication and through evolutions of complexity, subtlety, and ambiguity to today. Being a good communicator, individually, conferred benefits, and so the individual practised what they would say to others. If you really think about your thinking this should ring true. Thought is practised speech! In fact, often times in your interior monologue you may have asked yourself, 'who am I talking to?' (Speaking of which, when we talk to ourselves and say something like, 'come on, you know you can do better than that', who is talking if we are one coherent individual, as accepting the conscious 'I' defaults to?)

As Noam Chomsky alluded to recently, language must be stored in a hierarchical structure within the brain, and one might guess that the great neocortex development (the particularly human brain development) was very much in sync with refining the competitive advantage of communication. It may well require a synthesis of language understanding and neural mapping of synaptic firing during language to decipher the code of thought, but generally speaking it will not need to be done any number of individual times, as vast portions are almost certain to be identical/similar from brain to brain. Again, testing people with disorders and with a lack of comprehension of certain vocabulary, etc. may prove very helpful when conducting the physical representation of this neuronal activity. One strong hint at the centrality of language in consciousness may already exist in records of feral children's intellectual under-development.

Rather than be immediately dismayed at the vast array of words and sounds we make, and trying to give each one its own unique space in our brains, it is most likely much more useful and realistic to think in terms of the branching out of language; indeed, one can view language as this model of iterations of differences in any event, and imagine a model a bit closer to the tree-like structures of the neurons themselves. That is to say, there are categories of words' neural representations and then within each category there are divergences and associations, and so a grand blueprint can be imagined. It should be kept in mind that the neuronal capacity to change role should be evident in those who, through disability or other circumstances, can communicate by non-verbal means.

Just One Part

The neuronal patterns within the brain and nervous system are not just responding to our ponderings from moment to moment. In fact, as a part of your brain's system is rehearsing possible motor movements that would generate conversation with other people in your life, your brain is also regulating your breathing and your circulatory system as well as monitoring and informing systems such as digestive and lymphatic, secreting bile and saliva and a range of other activities. The element that we feel is our consciousness or the 'I' is just one other part of neural activity. Of course, it is a centrally important element, and does in many circumstances have executive decision making control within the dynamic, hierarchical network that is the human brain, leaving aside the Libet work mentioned above—'it' can to some extent decide on things that the majority of the body is cajoled into doing, such as 'taking a walk'. Also, it is tied with the language that has developed with our society's development and has placed eminence on being able to analyse and comprehend our environments. In a brain model of hierarchical networks it is likely that this 'element' sits towards the top of the hierarchy. Current computational neuroscience would support the idea of layers or a hierarchy of networks. This might mean that certain neurons taken together lead to a certain interpretation (let's call it interpretation 'a'), and 'a' taken as one particular outcome at a higher level may then join with other networks b, c, etc. that when taken together are components of another higher level interpretation.

It has evolved also, as have we, with a picture of us as a single entity that we must be spatially aware of and protect at all times. Being able to identify ourselves, differentiate ourselves from other entities (don't walk over cliff, steer bicycle through open gate), and a whole range of other core components of consciousness can be seen to have evolved with all other parts of the most complex informational pattern we know in the universe. But even here, again, we find that employing this conscious part too much can be unhelpful, as in when cycling through that narrow gap, we 'feel' our way as opposed to 'think' our way through much more easily. There are other activities such as playing music, sporting activities, and so on where we benefit from quietening the conscious 'I'.

When one outlines this 'non-magical' explanation of consciousness, there can be a worry as to such concepts as free will and criminal behaviour, for example. If we are not conscious (or consciousness is just a function of biological processes) what differentiates us from other creatures or objects? Should young children or adults with developmental difficulties be treated like dogs then? Of course, we can be aware of the scientific origins of our consciousness and aware of the errors evident in our selves' perspectives without abandoning vast tracts of our society's developments and histories — vast tracts of human wisdom accumulation that have none the less proven useful in our constructs. (Thankfully — though we often fail to see it — it is cooperation that has led to our relative dominance among many species, and this ability to cooperate is centred, in fact, on empathy — and so it is perfectly reasonable and logical to detest the idea of treating any child poorly. We feel others' pain. One might argue that recognising the lack of specialness about our own conscious thoughts will give us time to rethink the consciousness of other creatures, and perhaps we will also stop treating dogs as we do now.) This is akin to many situations in science in that we must inhabit the world that makes sense to us generally without denying the logic of non-human sensible discoveries of reason. As said, we do not walk around focused on relative motion in space-time nor do we worry about the empty nature of atoms as we look at the floor beneath us. We, in vast proportions, empathise and love children because this is how we evolved. We will penalise by law those who act of their own free will, because this makes human sense. Choice rooted in biology is still choice as it is one of the functions or outcomes of any complex structured entity's individuality.

***

I should mention one counter to this centralising of the role of the evolution of speech in human consciousness — that there is a consciousness without language, isn't there? I mean, an ape is conscious. It is aware of itself and its space and that it is a separate entity as it manoeuvres itself around its surrounds. Well, a flower leans towards the sun. Is it aware of itself? The question here is not when does this self-awareness occur, but is there any such thing as self-awareness? I am hoping that as we go through this chapter you might begin to come to see that the reason we have such a difficulty in pinpointing where consciousness is to be found in a progression through the species is that it is not to be found at all, really! The leaning flower, the swinging monkey, the thinking human, it's all the same: it's just that we as humans can't see it!

Just One Part — Part 2

Thought, as we know it, then, forms the interior voice that I use for 'my' thoughts. And we can claim, 'I think, therefore I am'. However, inherent in this is also the need to separate the 'I', referred to as the subjective voice, and the entirety of the underlying human structure, of which, in fact, this voice is just one element.

In fact we know that this subjective 'I' does not necessarily always have decision making authority and full control over our own bodily functions. Again, try telling your heart to stop beating! Or hurry up and grow, nails. We might say while walking down the road and listening to the birds and seeing the flowers; 'I pondered the economy'—yet all the while not only am *I* absorbing masses of sensory data and activating muscles and lung capacity to the walk; also my liver secretes bile, my skin cells are coming up through the layers towards dust, and my digestive system moves my waste a little further along—these are all parts of what 'I', as in the totality of the individual unit of 'me', am doing. 'I' am not pondering the economy while my liver secretes bile. I am doing both. My liver particularly is secreting bile and the elements of my neural network concerned with language internalisation in conjunction with areas concerned with absorption and retention of all that I have encountered in the area of 'economics', or my associated memories, is/are pondering economics (I am doing all this as an evolutionary hangover of the benefits of communicating well, and so rehearsing what I might say to someone on the issue!).

Returning to the concept of 'rehearsal' may well lead us to believe that we are in control again, or that I can ponder what I will say in a month's time at a meeting, but again this leads us to what the 'I' is that gets to do this planning. In fact, while we do rehearse as well as act out these vast arrays of muscles, organs, and systems with the immensely sensitive manipulation allowed by the vast numbers of neurons involved, there is still no planning. There is only reaction and prediction.

When something forceful impacts a wall, it doesn't think about what happened—it reacts. Perhaps, if it is built in a particular way, it collapses, or maybe just shudders a little, but it doesn't think about it, it just happens. The human system's reaction, again, as with the wall, is certainly not planned; it occurs. It is registered by the human entity via a range of systems that we collectively call our senses, and information is fed through to what we see as this separate information processing unit, or the brain. The brain certainly processes information and, again as with the wall, it reacts. What I would like to argue here is that there is no substantial difference between the wall's reaction and our brain's, other than of what we perceive as complexity. We have argued already that the voice or the 'I' is not some omniscient, omnipotent, or even all-encompassing entity, but one set of nerves connecting or almost connecting to muscles when rehearsing or acting for thought or speech. Of course, it is a part of a vast network that is absorbing information, controlling movement, and referring (via its own inherited design) to an internal model of the external environment continuously. These vast networks are developing from infancy, indeed, even in utero. Like similar plants differing due to the weather and soil there is the initial genetic design plus the plasticity that allows the brain to learn from experience. The initial drivers of this learning can be seen with the simple cries, smiles, and laughter of a new-born baby that needs feeding and reassuring warmth and comfort from being held. It imitates and pleases carers and has positive reinforcing chemical connections to this feeding. It is all the while producing new neurons that lay down layer upon layer into networks of different functional outputs, again as dictated by the genetic blueprint plus with experiential-based particular wiring.

Evolve

What are we really then? If you would, do the following in your mind's eye.

We do have an inbuilt horror of our own innards. This is entirely sensible as it is a major difficulty when they see the light of day for our entire entity's integrity and survival!

But, nevertheless, there are many images we have encountered that might allow us to peer, imaginatively, inside this entity. Again, this is useful in dispelling the subjectivity trap — perhaps begin by peeling your skin and outer fat back so you can see the interior; as you travel along with sugars and salts and take the grand tour, you see something founded on 200 cell types leaning on 206 bones, with a muscular system constructed of the same material as the bones but in a much less dense fibrous state, and so allowing greater flexibility of movement while the bone protects against force and maintains stature. We see a system of organs that to a large extent support the maintenance of the entity by allowing the ingestion of these salts and sugars and fats and so on so they are broken down and have their useful elements extracted by the engine houses of cells (mitochondria), to replace old cells, to allow blood and oxygen to be carried around the body largely through the circulatory system and allow the energies ultimately to be used up in movement control over the body.

These fibres and muscles are continuously stretched and contracted in a very fine manner (though nothing comparable to a bird's flight or other dexterous creatures' activities, but what seems pretty clear is that this is our design). Many of our organs act as pumps for various ingredients we wish to carry around to their needed destinations, but many also act as a complex disposal mechanism for the materials we ingest that we do not use for these energy consuming purposes, our waste management system. It truly is a very central component of our entity and does fly in the face of our self-importance a little, where we see this foul smelling disposal as almost 'beneath' us and our important thoughts. Though, surely at some point in your life as your 'I' was making an impression, your waste management system exerted its greater influence, forced you to make an unseemly rush to a toilet, and showed you who is your boss!

But in all of this does it not seem very unlikely that something that sits atop all else, so much so in fact that it is made differently to the rest of the universe and might be designed by a god, or an intelligent designer, would put so much influence on waste? Why not design us or the foodstuff so that it just contained what was required? By the way, in our lofty self-explanations where does sleep fit? Is this not much more a reflection of our ordinary physicality and energy supply and usage (particularly the neural centred work—this awesome processing of sensory information, absorption, and reaction) than of the mind-spirit-soul analysis?

Is it not so much more believable that our entire entity is the outcome thus far of our particular species' evolution? As such, can we not see that the brain orchestrates so much of all of this, as the neuronal network absorbs a vast amount of sensory detail, but also sends an enormous amount of reactive pre-programmed instruction around the body? It is, of course, not just what we or our 'I's feel aware of, but also all those chemicals that we carry around our entity serving a range of support functions in themselves, plus the digestive system doing its thing in the background from your 'I', and much more, are all originated in the brain. How, though? Because neurons, ultimately, emerge from their large networks to nerves that activate or contract muscles, muscles which control motion as in walking but also as in blinking, swallowing, seeing, digesting, releasing chemicals, controlling vocal cords, childhood imitation and rehearsal of speech, and so on. It is literally the nerve centre. We can say, then, that to know what a neuron does you must follow it. Neuron is as neuron does!

Increasingly accurate imaging should, as such, show massive overlap between saying something and thinking it.

The Hardest Thing

But surely there is an 'I' who gets to decide what the 'I' thinks about, yes? Well, and this is the hardest part of this whole thing... No! There is an illustrative line in the film *Inception* where one character asks another if she ever noticed how in a dream you just are somewhere, but you never know how you got there. This applies for your whole life, also. You don't remember how it began. It seems on a minute by minute level that there is an 'I' who dictates everything, and practically of course there is a sense of that, but stepping outside your own mind for a moment and looking at the humans around you; their systems have been firing since neuronal development in the womb, systems that are affected by interaction and design and much, much more than what 'we' currently perceive to be the key learning method—the spoken/written word; which, in many ways, is providing just a poor summary of the system interacting and reacting with its environment and according to the initial programme and with the plasticity to learn. *Central to this system is a communicative function that claims ownership of the whole entity, centralises itself in an exaggerated fashion, and claims the* ability to start its own 'thought' processes creatively in any fashion it so decides!

Shutting Down a Spirit

We might also point out that we even know that this system can cease operating and be rebooted. Certainly in an anaesthetic or coma recovery there are strong examples of how robust our neural system is. I will mention this in Book 3 towards the very end also, in what I believe to be a very important ramification of this whole line of thinking. I would like to highlight another stronger form of this though, it's called Profound Hypothermia and Circulatory Arrest. This procedure is used when treating difficult to reach aneurisms. What is this procedure? Well, the body is cooled to ten degrees and all communications between neurons effectively cease. By most definitions the person is dead. Believe it or not, the person is reheated and rebooted once the operation is completed.

I encountered this procedure from a short story called *Killed by Bad Philosophy* by Harvard neuroscientist Kenneth Hayworth. This is, as Hayworth points out, an effective death knell to so many anti-physical theories of the brain, and, I feel, a very clear example of how this thing we deal with so delicately is quite a bit more robust than we think. We often flood it with chemicals in the course of a Friday night! Many arguments that scream that the mind or 'my mind' is such a never-changing entity — that we are always who we are… this is just fundamentally not true; in the subjectivity trap we hold on to the notion that I am me and I always have been, but, in truth, we change all the time, and change profoundly over time. There is a coherent pattern in our DNA, but there is great robustness built into the system. As I say, on Saturday morning after the Friday night of neuronal assault through substance abuse we may well state, 'I don't feel like myself', but we don't take this literally. When the 'system' is rebooted from Profound Hypothermia and Circulatory Arrest or, indeed, when it comes to from an anaesthetic, we in some respects are just turned on and find ourselves mid-thought. That is to say there isn't a point before we are thinking that we (some central Cartesian figure) think about what we will begin to think about—the light just comes back on and the engine turns over.

Neuron Is As Neuron Does

Really, could it be that easy? Well, it most certainly is not easy. The numbers involved and the complexity is still 'mind-blowing', but the answer is yes; I mean, what else could it be? Neuron is as neuron does. The neuron may be excitatory or inhibitory or it may be at the end of a nerve and sense environmental effects or affect the contraction or relaxation of muscles. Of course, there is the ability to have immense complexity arise out of the species and individual learning that occurs within those parameters. Undoubtedly there are layers of interpretation in the brain's system. In other words, a particular neuron will spike if the skin is touched by a particular element, say water. There may well be millions of others conveying the same message or reinforcing it, and the retina is picking up its own light patterns and (using derivations from what is learnt to mean certain things over time) concluding that it is raining. There may also be a continuous loop of thought along the lines of some existential angst. This is embedded in the rehearsal of the vocal cord and diaphragm controlled exhalations of air that by control signal certain learned and shared patterns among peers. So these pathways of neuronal release of chemicals are very much inbuilt, but the plasticity and, particularly in early brain development, the major re-wiring (or cessation of connections) that occur in everyone all allow for a variety of interpretations by the learning child brain.

So you have neurons firing to deliver information about the environment, internal and external—and to an extent the individual neuron is contributing to a number of outputs. So, when we have nerves in the brain in, say, the medulla oblongata and in the central nervous system and the periphery all activated by a somatoreceptor on the skin, say of the nociception variety (or pain receptor), that are activated, there is a response that is instantaneous. Motor pathways-led withdrawal of the affected area is the normal option here. The hand is removed instantly from the hot surface, for example. But in reality, this is all that happens throughout the brain. As difficult as this may seem, even when we feel we are sitting and planning, we are merely reacting to learned patterns of circumstantially suitable paradigms of behaviour accounted for through the communicative function's seeming dominance of much of our entity and a whole range or multitude of other concurrent processes. It is exhausting, perhaps, to even try to grasp a moment of all of the work under the surface; again, this is why the subjective 'I' provides only a summary. To repeat—our conscious 'I' does not see the components of itself (the workings of neurons at each instant), it summarises activity in a useful way and as the result of the communicative function evolving to allow for this.

Different brains—with differences in structure, erhaps some of which are innate and some that are learned—may 'see' something differently to another's perception. What makes it right or wrong? Up to us! Mostly, right and wrong must be about common consensus, as to claim there is a universal anything is falling back into bad habits. But, look at someone with Anton-Babinski syndrome (where people who are officially blind are convinced, and remain convinced despite whatever evidence you and they provide (like walking into things, etc.), that they can see). So are they wrong? 'Of course', you might say. But that is under the assumption that we sighted people are right, in some *universal sense*.

Undoubtedly what we (normally sighted people) portray is more useful when it comes to avoiding injury, but do we see the almost total emptiness of the atomic world that underlies everything within and without us, or even the full spectrum of light through which we claim to see everything? No. So you learn and react. Rather than saying, then, that one person sees and the other is blind, we must look at two creatures with different functionality. Again, there is usefulness to one's interaction with its environment and movement and energy absorption, etc., but one plant may have a different root pattern or a flower or a slightly more efficient pollination system in some circumstances than another. There are only differences; the judgement of better worse, etc. is a human-only contribution.

This is not easy going, is it; this attempting to close the so-called 'explanatory gap' by highlighting the complexity of the mechanisms involved? It needs restating and revisiting to shake something so fundamental.

The Subjectivity Trap — Again

How then, is it that consciousness is a mirage?

Let's look at how I arrive at the ability to say 'I am conscious'. Clearly I could not make such a statement at six months of age. The language and the meaning of words are absorbed over time through my neural network; more specifically, my auditory system is attuned to certain enunciations through highly sensitive sensory neurons. I can phonetically mimic the sound of 'con-shus-ness' and then return it via the motor neuron control of muscle movement leading to the diaphragm, vocal cord pressures, mouth shaping, etc., until I can hear someone say I am conscious and also say so myself. In fact, I can just think it. By thinking it I am showing a further skill of humanity within the broader evolution of communication and that is as follows: to not say something, to prepare it and try to assess what sort of reaction it would receive among my peers as my ancestors would have dealt with particular gestures and basic primal sounds among their tribal grouping. *But, how do I know what consciousness means and when to say it? Again, this is very difficult to understand but not impossible, it involves watching a child very, very carefully for a very, very long time; we will initially see just cries and instinctive movements of bodies, we will see the role of 'mirror neurons' and imitation, we can imagine the ingratiation of food and learned impact of pain, we can see the associations of sounds, the errors that we all find so amusing in this learning child... and, in time, we will say something like 'aww isn't he/she developing a great little personality'. If we look very hard we see that the child arrives at the understanding the adult has by an iterative process — a simple programme is laid down with early words and obvious sensory connections and over time more and more differentials are applied to this, and most words are for a while misinterpreted; but we build again and again and are corrected — and all of this leaves us with this wide vocabulary that we feel cannot be explained by the simple imitation and the iterative adding of differentials towards complexity. But it is purely interaction and pre-programmed reaction.*

Returning to our earlier story and our tribal forefather — to receive more credit within their tribal society an individual ancestor had, as with all others, benefits to be gained by increasing the complexity of messages. As with other motor functions this is a rehearsal; something we now know happens with all motion in the supplementary motor cortex. From cave paintings of likenesses to symbolic representations, from primal common screams and shouts to nuanced language usage, we can surmise that language was both a tool and contributor to the increasing complexity of our cooperation, our social interaction and development. Why, though, is the interior monologue so obsessed with 'I'?

The ability to communicate was inherited through species evolution, but of course, as with a bird's ability to fly, still sits within the singular entity of an individual human and, as such, has a starting point and centrism of 'I', as the communicator tried, largely, to express their own messages to their own benefits throughout the evolution.

If I could second-guess some form of objection here along the lines of 'who or what are you referring to when you say "I" or "they"? Surely we must concede that there is an actor — and that this defines consciousness — this sense of a self?' For the purpose of continuing here I would like you to liken the actor at its most fundamental level to a flower that leans towards the sun. It is an individual, self-contained agent — it chooses to align itself in its growth, i.e. it would align itself differently if planted elsewhere — and I know you say 'but that is just a biological cause and effect, a result of chemistry'. Well, this is what we are also. Again, don't despair of free will please — it exists; it is just that it isn't magical and unearthly but is an element of how we lean!

The awakening of this communicating body came with severe perceptual constraints, including, for example: the concept that humans are special and have gods that look down on them, Earth is the centre of the universe, and the human ability to communicate and to rehearse that communication is completely unlike any other functions observable amongst Earth's creatures (even allowing that some other creatures also have some communication skills, which instead of saying does not make the ability to communicate, perhaps, a little less important, rather we say it endows some other creatures with these near magical, human attributes!). Within the subjectivity trap, it is better to explain our 'selves' by assuming that humans have a universe of unfathomable proportions designed for the purpose of their existence, or that a deity of no substantiation generated these mystical creatures—us, humans, or at the least human functionality, is so different from all others that it comes from a totally different source.

To Find Out What a Neuron Does, Follow It!

I believe in this context the neuronal circuitry we view today is still awesomely complex and operating with a wonderful range of potential (at least from my human perspective), but the task is not as difficult as we assume. As I proposed earlier, to learn what a neuron (any neuron) does, we must simply follow it. But its impact on different networks and hierarchies will need measuring. One more thing we may have going for us, in trying to unravel the brain's functions back to its neuronal manifestations, is that, although undoubtedly complex, our ally evolution will see an underlying simplicity at play!

As said already, neurons can largely be sensory, motor, and inhibitory, and while it is difficult to concisely describe the almost endless potential combinations and pursuant actions, it is not quite as difficult as first glances might suggest. While there are a number of different signalling chemicals—and indeed each neuron is somewhat unique and the role of the glial cell seems now to include repair and support as opposed to the afore believed sheathing, and many more unknowns—there is a huge amount of commonality across neurons: while they form massive complex systems and networks and have the ability to play different roles in many tasks (and also, importantly, the

Chapter Five: What is the Subjectivity Trap?

ability to change their roles and the tasks by aforementioned Hebbian learning evidenced in synaptic plasticity), we should celebrate the commonality of purpose to interpret stimuli and to react via control of the varying parts of the human anatomy. (Or, one might stay, stuff comes in to the human system through the nervous system relays, and stuff goes out in actions and rehearsals of actions…)

There are also large amounts of commonality across the human anatomy, with mostly all body parts stemming from quite similar fibrous components and following elements of the unique master code (or DNA) of the individual. Very interestingly, lately, the brain seems to be revealing that it unfolds into grid-like sub-structures. As difficult as it is to see this when we view the messy porridge-like substance, when it is by computer imaging 'flattened', it seems to be like grids in its basic design.

The circuitry at a given time may be likened to a city of ants, where each neuron may be likened to an individual ant following certain tasks or functionality without any specific 'thought' attributed to the individual ant or neuron; it merely acts. On a grander scale, too, we merely act. Again, if nerves on our skin relay information that what we are touching is 100°C, then we appreciate that from our original DNA-generated circuitry a vast array of neurons will kick in their excitatory functions to generate movement in the arm and hand away from the hot surface. This is certainly at the simpler end of our actions. We can propose that the numbers and types of neurons employed take precedence here and, again, this will come from initial design (DNA and early development) and then Hebbian learning—an individual in some hypothetical case may have been exposed to repeated experiments where, after a brief burning pain, a stronger benefit is made available—perhaps the pain precedes a neurochemical release similar to many drugs that create euphoria and are injected by addicts. They might develop a tolerance to the heat, and not instantly withdraw.

Neuroscience experiments where an electrode is planted on one particular neuron shows that this neuron (and likely others in the immediate vicinity) spike or conduct an electrical signal if a particular shape, say rectangle, is at a particular angle, say 45° to the left of horizontal. If it is closer than this to horizontal or vertical, then the neuron will not spike. This is astoundingly precise. If the rectangular black shape on white paper is presented to you and it is slanted at 45°, or near it, this particular neuron will spike — if it is vertical or horizontal or, indeed, somewhere else on the sphere of possibility, it will not! If you watch a bird's feather react to miniscule variations in wind streams it is equally impressive. But why is this neuron spiking in any event? Well, it is feeding into an excitatory-inhibitory based vast network of sensory responses — it is merely responding to impressions made on the nerves of the retina by certain light patterns. It is a network that is much bigger than the flower's locating of the Sun, but not 'philosophically' different. In evolutionary terms it originates in this long story of chemical soup, through to creatures stumbling through the filter of natural selection, and then emerging as pattern recognisers able to avoid dangers, seek rewards, and so on.

We can conjecture that what informs the early neural architecture is genetic and natural selection prioritisation plus Bayesian probabilities. There is a hierarchy of neuronal activity as well as the predictive Bayesian model. We now know also that a part of the brain's function — perhaps its entire program — operates by means of this prediction; initial sensations are drawn from a memory bank of similar synaptic connectivity, as when we hear the first notes of a tune. (We will revisit the musical reference very soon.) As we run through the initial notes we open up a lot of possible songs from our memory bank, as the second and third notes emerge we converge until we happily immerse ourselves in the melodic familiarity and can almost sense the predictive model when we can pre-empt the words of an old song that we would have little or no chance of writing down without the musical prompts and the re-running of the neuronal pathways that emerge.

Perhaps added to this the reader should read about Jeff Hawkins' memory-prediction framework model. Utilising Bayesian probabilities I would argue we should also include distributed intelligence (entirely distributed) and hierarchies of significance in our current models of broad functions of the brain. I wish to briefly consider Hawkins work now and it should tie in with a piece on music later on.

The Memory-Prediction Framework

Research shows that the neural system takes sensory information and extrapolates forwards to a model of what it expects. And so, as highlighted already, visually speaking we do not bother to allow every retinal registration make its way through the sensory network prior to arriving at a 'full picture'—we make predictions of what the remainder of the picture will be from a very early point. We can see how this fits with circumstances where there is more effort required (or concentration) in absorbing new information and 'surprising' information. Our brain's energy usage increases and we are 'on alert', if you will, when there is a surprising or unfamiliar component in the picture—again, sensible enough from the core survival instincts evolutionary path. What you know you feel safer with—and you must try to make decisions fast while watching particularly for abnormalities.

This is an interesting perspective, as it fits with the simple neural model we know, i.e. Hebbian learning via repetition of synaptic firing between neurons and the strengthening of the synaptic connections; as stated above with Eric Kandel's work as example, the familiar becomes activated a little easier—the brain looks to establish patterns that allow recognition. What this may suggest, and goes to the core of this piece and the entrapment of subjectivity, is that the brain's workings are, in fact, 'unthinking', in an objective sense—and so we may have perceived ourselves through the mirror and discovered that we are at once hollow and brilliant, at least 'brilliant' from our own viewpoint. Our perception is something we really should perceive positively. This certainly is an extension of the decentralisation that began with the re-evaluation of the Earth as the centre of the universe.

Bayesian probability theory fits perfectly with the memory-predictive model. This argues that neuron networks work on probabilities and experience and seek out the most likely previous experience with which to compare and react (think of music recognition and so on). So when the skin senses rain, a vast set of sensory neurons are activated from all around the body's exposed skin—the numbers in terms of texture and size and frequency are weighed in this Bayesian predictive model and the information is encoded in the neuronal pathway that is activated. When one is thinking to oneself then I would argue that this is an internal loop, and one word does lead to another; although I will concede this can be very disconcerting when thought about.

Animals

Look into your pet's eyes—talk to him or her—is it a dog, a cat, a hamster? Tell it something, use its name, ask yourself; really… is this creature not conscious? Imagine the pet dying—is it then unconscious? Can you now see that consciousness resides in your pet without imagining this awful tragedy? You may concede that Rover the dog is conscious, but then dogs are very smart animals. OK, but what about a pig? A pig is even smarter than a dog—but your eye isn't as experienced at looking into a pig's eyes for consciousness. What about a dolphin, a shark, a salmon, a goldfish, an amoeba? You get my point, where does consciousness—this special thing—begin?

Let's take another look at this misconception of the human/animal divide on this mysterious consciousness. Think of when we see a glazed-eyed gazelle run from a tiger. Do you think that the gazelle is conscious, not alive but 'conscious' in the human sense? Well, it is undoubtedly reacting of its own accord. It is in the midst of a fight-or-flight bolt of adrenaline, its central nervous system is coordinating its rapid motions and turns. I may assume that one argument offered here by phenomenologists is that the gazelle cannot contemplate what is happening to it. Are we really sure about this though? Do we not accept the unity of entity and the awareness of the event—is it perhaps that the gazelle can't (with any great sophistication) tell any other of the event, but that 'it' is aware of what is happening? When we are contemplating, is it at all possible that this is a residue from the evolutionary development of the benefits of human communication? In other words, if the tiger was chasing us, we would still have a little internal monologue running at least intermittently (or when the more 'primitive' parts of the brain and panic had not taken complete priority)—is it possible that this is because there was a benefit to a member of a tribe by being able to describe the beast and event and means of escape and so on, and so we see the evolutionary benefit of this 'talking', and, so, even though no one is around, we still 'talk' to ourselves…?

The notion that I merely react, and the thoughts that I have stem from the evolved ability to communicate coupled with the internal representation of the external world that is being closely studied and understood all the time—as well as the predictive and evolved loops from previous thoughts (or unspoken words) plus the prioritisation of being able to 'represent' events from the 'hunt' to our tribal members, and so concentrating vast portions of our energy-sapping brain to this speech and practising of speech—to repeat—should not precipitate societal revolutions where concepts such as free will et cetera, are reconsidered. All that needs to be acknowledged is that we are human and not special; of course, it would still be pertinent to view all that we have collectively and historically constructed within the 'reality' of being human as essential.

The background context is all that changes in that some unsupported, universal pre-eminence of the 'consciousness' world is not retained. As with relativity or many other new interpretations, we will continue viewing time and space as predictable and singular entities because, as humans, that is what we can relate to—but we should not deny the logic of that which we don't experience directly, and so, as said, we accept relativity, but drive our cars without reflecting on it.

Similarly, it would be imperative that, while we appreciate the true lack of a special place for us, and the predictive Bayesian model of our Hebbian learning brains, we do not discard the legal frameworks we all now understand and largely obey to society's benefit. Healthily, though, the wrong and right of things are discarded in favour of the consensus of cooperation. People with abnormal brain functions are truly understood as different, outside the norm, but the judgment of right or wrong or such inane concepts as evil return to their origin of ignorance. The other undeniable result of seeing ourselves in as non-egocentric a way as possible is, almost counter-intuitively, the almost limitless potential we see we have. For as interactive units of matter we develop in our human way the ability to manipulate our environments with a consistently growing, shared knowledge all of the time with only time and application limiting the possibilities.

Perhaps taking a break to look at the role of perception and its dubious position will help at this point.

The Role of Perception

If you saw a photo of your author, and compared it to one, say, of George Clooney, well, suffice to say you would note slight differences!

This would be a difference, in fact, we would see incredibly clearly. But, in truth, look at the physicality of the differences between people who seem to be 'ugly' and those that seem to be 'beautiful'. The other guy's nose may have been 0.75 inches longer than the first or had a gap of 1.5 centimetres between his front teeth, etc. We could compare the two of them side by side, with one a full four inches shorter than the other, again massive differences… but what if we introduced an elephant alongside this pair, or a whale or a tyrannosaurus rex or a crab or a planet, etc.?

It is not surprising that, being human, we concentrate on human differences and identification like this through relatively minor differences (on a cosmic or even planetary scale), but this bias undoubtedly skews our thinking towards the differences pretty drastically. The 99.9% similarity of the genome, however entirely valid that remains as a meme into the future, further illustrates my point. I think it is important to sit back and try as best as we might to stop thinking like a human, or, at least, look to accepting that we are bound by this 'human' perspective, and be open to its necessary biases and weaknesses.

More on Perception – Perspective

Here is an example of the role of perspective. Instead of merely saying I was born, we could point out that without an incident during a war when my grandfather narrowly escaped death through the assistance of an enemy, or an illness that my great-great grandfather almost died from in childhood, or a large rock that just missed the head of a distant primate ancestor, or a comet that missed Earth a billion years ago, and so on—with all of these innumerable near misses I would not have lived; and so I am incredibly, no, ridiculously, lucky to be alive. Or we could just say that my parents had intercourse and I resulted. This is an interesting contrast of perspective that seems to be at the core of many of our existential musings. This is indicative of the human-centric perspective that we must have, but, yet, continue to deny! It is useful, and impossible to go without, undoubtedly… I just think we shouldn't allow it too much pre-eminence. Imagine a football match with the team you support and their arch rivals. There are two identical penalty claims for either side. Your perception is likely that the one for your team is a penalty and the other is not. (If football doesn't do it for you there are many lifetimes of comparable examples of the fact that we are walking prejudices.) Why? Because we are subjective. Why? Because we can't be anything else!

There is a gentleman that resides in a psychiatric unit who stands at the window all day and moves the Sun with his mind. I am willing to bet you found that somewhat amusing. I know I did. That was my reaction, it is so ridiculous. It is clearly very wrong; but we should really try not to laugh, because everything we do is affected by our subjectivity and our bequeathing a universal or cosmic significance to ourselves that is just unwarranted.

In fact, touch a nearby table. Be sure you touched it now. Well, just to show the ubiquitous interference of the brain's inference, in a way you have not actually touched the table. In the useful sense of how you have always imagined touching a table to be, you have, of course; but think about the sensory signals that spike and travel to your brain and are mimicked in the storytelling narrative, rehearsed speech function to allow you say or think—yes, I have touched it. If any of these components malfunctioned you wouldn't be able to say that you 'touched' the table. So you rap your hand hard on the table— you are sure that you can feel it now, as you can feel a pain from hitting your hand on something solid. Well, in contrast, we could take the example of the very tragic condition of congenital analgesia, where patients do not 'feel' pain. Many of these persons will touch fire, walk into and cut themselves on sharp objects, and many will die in childhood as a result. The hand, or whatever body part is coming into contact with the sharp instrument, may bleed and you can see the blood loss, but the brain is not processing this information in the normal way. Of course, also, physically speaking, your hand does not 'touch' the table. The atomic fluctuations that are the centre of these two objects are based on repellent forces and large tracts of atomic emptiness. So even without looking at congenital analgesia, in the microscopic reality things don't 'touch' each other and so your hand, physically speaking, does not touch the table!

Let us not laugh too hard at the man that moves the sun, but appreciate that all we all do is create an internal virtual reality, and those we see as doing it incorrectly are really, in as objective (or 'less wrong') an analysis as we can muster, just doing it differently.

The Paradox of Fermi's Paradox

I am guessing you have heard of Fermi's paradox, named after Enrico Fermi, Italian physicist, who first ruminated on how there could be so many habitable planets and such a long universal history and, so far, no sign of life 'out there'. With every passing week lately it seems a paradox that is becoming ever more difficult to account for, as we now claim upwards of 60 billion habitable planets, just in our galaxy.

This is missing one absolutely essential point that again looks at our biased perception. It assumes that planets which develop life, eventually, develop a life form that wishes to communicate with creatures not from that planet. This seems to be seen as a conclusion to the evolutionary process.

In fact, let's look at how we view evolution. We who are one species of approximately one billion through Earth's history decide that we are the conclusion of evolution as a starting point, and that there are other co-existing species that aren't as evolved as us. Now, fair enough; there are very many simple organisms that seemed to have developed towards more complex organisms, and I can see how you might make an argument that a lot of that one billion were inevitably going to develop into more complex creatures and would do so anywhere else. Just two points on this: there have been a number of near total extinction events in Earth's history (most famously at the time of the dinosaurs, perhaps) and it is very difficult to see the inevitability of humans in this context, when that cosmic event changed the evolutionary tale so fundamentally. Also, if we can see ourselves, absent of the homo-centrism, as just one species — one that can and does communicate beyond our planetary scope — there are over a million species on Earth right now, many of them existing in much greater numbers than us also. So, being as objective as we can, is it anywhere even nearly likely that 'we' or another communicative species would be the inevitable evolutionary output of similar planets existing in the galaxy and universe? And so while 'life' almost certainly does exist in many forms — it is entirely logical that there are no 'communications'.

***

I don't doubt that this is all extremely difficult to let go of. It is always such as our internal world representation attempts a significant redraw. We rely (don't walk over the cliff or hug the strange slithery creature) on certainties of what we know and caution against accepting what we do not. And so a challenge to a fundamental 'belief' as part of our internal representation is very difficult to incorporate through our communicative function into our reaction to events at the best of times. So, anything that touches on the notion of how these redraws, or indeed, the internal world representations themselves, may be falsely based is most certain to cause discomfort and have you mentally leaning towards rejection. Of course, many may be introduced to a work like this and begin by parking the little amount of 'less wrong' or non-specific prejudice thinking that they can muster, and have a starting point or argumentation where they say to themselves on beginning, 'this is crap — this is all just crap — I don't really care what is being said — it is all just crap!' I can only ask that you hang on all the time to what can be reasonably assumed is the most logic we can muster. The aim here is to not take an intellectual shortcut, not to assume anything easily, and not to move your mind because of discomfort.

Just think through yourself about the vastness and speed of light and space, and the smallest grain of sand and how we can make it smaller and smaller and smaller, and how we have approached everything here with great care to not make unwarranted assumptions, but to draw on less wrong conclusions, when errors seemed obvious. As I said; you can trust, I hope, the author is being sincere when saying that he did not have an agenda but only a curiosity when travelling these mental journeys, or re-evaluations. I find it useful when building up this appreciation to look at instances where we, the entrapped 'I', can glimpse behind the screen, if you will. There are so many — once you become adept at observing — such as outputs of the brain that seem to straddle what is normally within and outside of the subjectivity trap, and wherein we find dreams and hypnosis, as well as a myriad of exceptional behavioural cases and disorders, accidents and diseases, that all show us the mechanics of our brains and the lack of anything divine about them.

Chapter Six
Ways to See Beyond the Subjectivity Trap

Running

Here's a really cool and simple exercise to bring yourself a little bit closer to your own illusion. Go for a run, or some form of exertion activity. As you start out on this run, think to yourself about something complex, something difficult to figure out. Be sure as you start your run or activity to keep thinking of this thing, then increase your speed and see where your thought has gone. But, for the really good bit, do it again. Tell yourself this time as you warm up that you will continue to think about this problem, be it mathematical or whatever, be certain and convinced, this is your mind you have control of it — it is you, in fact, how can *you* not do what *you* want? Now increase the speed again, until you are going flat out. Note, particularly, how your 'I' will try to give you an excuse, or tell you that you will think about it later, and so on. At this moment, you are of course teeing up a good insight into the illusion, because you are trying to maintain something you feel is always completely in your control, but your biomechanics (which you see as a separate thing) is just plum running out of oxygen and can't afford to be pumping blood to your neurons.

Innards Gazing!

We looked at this a little before, but the thought experiment above with the peeling back of your outer layers and looking into your body helps. Can I also add that you might imagine sections of it starting to malfunction or come away from each other, and try to use this to see its interconnectivity and mechanics. There is a theory of universal death which projects entropy's success and expansion of the universe until everything pulls apart from everything else (including atoms) — 'the big rip' is what they call it. Using this image of the neurons pulling apart and the accompanying loss of function and linked into diseases of the brain and the role of trauma from force or stroke or the role of plaque-like build-ups and the breakdown of communication between neurons and fields of neurons may all help to get this mechanistic image embedded against the normal 'I's response, which is protective of even the imagined integrity of the brain.

Meditate — Transcend — Turn Off That Voice

This constant reporter that we have evolved is in itself a source of a lot of mental activity and so drains on the body's resources (again, note how our mind's separation from our body seems rebuked by the clear link between mental productivity and a lack of nutrition/sleep, and so on). It is pleasing to us if we can quieten this a little during our waking hours. In many ways as we push through the body movements and breathing of so many exercises such as yoga and other forms of meditation (indeed the chant-like nature and prostrating oneself in prayer) we are looking at methods of quieting the inner voice.

This is seen in so many walks of life. The silent, incense-filled, large, darkened space of a church, or the dressed down, floor-centred mat for a yoga class, the sound-waved rhythms and shouting of a spinning class, the intense 'physicality' of sex, and so on, leading to, in some cases, a quietening of the interior monologue, or narrator, or communicative cognitive function through absorption of brain activity in other non-communicative activities — and considering the break in resources, is it any wonder that we feel refreshed and sometimes go so far as to take a feeling of awareness through senses, less the voice, as 'transcendental'?

In an attempt to give your neural networks a break from over-concentrating on this communicative function that is your 'I' (which is essentially what most people work on all day now) you can even try something as easy as listening. You may close your eyes and begin to concentrate on sounds, as you will pick up many that your concentration has heretofore ignored. This refocus will improve your breathing and make you feel more sedate; if you have been stressed (this is a bodily response where the system feels it is under threat and spends resources in sending adrenaline and cortisol around the body in preparation for difficulty — or in evolutionary reality, attack from a wild beast) you may well find that active listening is one of the cheaper ways to activate the sympathetic autonomic nervous system.

Just to point out, the subjectivity trap makes a number of claims on existing theories which I hope are becoming clear — the idea that the nervous system and the autonomic nervous system are different systems is erroneous, that is not to say that there are portions of the body that another part of the body (the neural networks we see as the subjective 'I') cannot influence greatly, of course there are. Another small one, of course, is instead of saying mind–body duality or the physicality of certain mental elements, or mind over matter or the body and mind are inextricably linked (which unbelievably is seen as a progressive opinion sometimes), there is no such thing as mind and body. There is only YOU, in totality.

Levitate

I appreciate that when anyone reads any part of many of my arguments (because they will inevitably be wrong if taken in particular contexts or if an additional layer of meaning is put upon them and then questions are asked of this layer) it is easy to reinforce the reader's scepticism and distrust; but, can I ask you for a moment to do that learned trick we have all seen from movies and the like, where we imagine our view hovering up above our bodies… OK, but now continue upwards, recall Superman when he could hear everything that was being said by everyone below, hear them discuss the arguments that these humans make that they are conscious, that they can control their thoughts, hear them discuss other things in life and love and career and thoughts and a lifetime of discussions, but move up further away into the truly vast silence of space, and see all of Earth's species doing their thing, see little creatures burrowing, see insects swarming, birds flying, humans talking, and all of them evacuating their digestive tracts' holdings; see the size of the Earth; and eventually it may be apparent that we are just of the universe (not that we 'are' the universe), and are in a position to create so much with true appreciations of our potential as well as our cosmic insignificance.

Another Clue

In body dysmorphic disorder people obsess negatively about how they look. People will feel themselves 'ugly' and will often concentrate on a particular blemish and greatly exaggerate its significance relative to the common view held of their image among others. People can have long running battles with mirrors that may become disruptive to their life's normalcy and can in extreme cases cause suicide. Schizophrenia is seen currently as a disease with a 50–80% heritable component and with a complex array of genes and environmental influencers assumed to be involved. It is also seen to have typical topological alterations in brain structure with increased mass in certain areas and decreased in others.

Essentially, it is believed now by most neurologists that these sorts of conditions are caused by a problem of connections. The connections in the neural networks or arrays of neurons synaptically linked to other arrays in given activities vary from person to person, and the differences are significantly more apparent when viewed in some extreme medical conditions—this 'connectomics' we will come back to soon. *I would like to state here that as we have proposed that all consciousness is really a virtual and artificial projection to some extent, in many ways it is 'useful' for our species when falling within certain genetic parameters. Or, to put it another way, it really is just that schizophrenia is* another way of interpreting surrounds, and it is less useful to the organism than more 'standard' approaches. But, to be clear, the idea that there is a right mind or a 'sane' mind is erroneous. It is merely that each mind wishes to be the right way, and we then embellish what is common with some form of 'right'ness — and by extension view differences from this 'normal' benchmark as 'wrong'.

I want to return to this concept of networking and connectomics very soon, but first an easy one of these subjectivity trap indicators I feel is **MUSIC**.

MUSIC

I happened to be watching, recently, a replay of Queen playing at Live Aid. Freddie Mercury sits at the piano, and after a little warm up he plays between three and four notes. Instantly tens of thousands of people roar in acclaim of Bohemian Rhapsody, but, importantly, to acclaim it, they needed to recognise it. I will return to how this recognition occurs in a moment. As they belt through their notes, what are the tens of thousands of people doing? Well, they are now jumping up and down, screaming in a near euphoric manner. They are in very close contact with one another, stood in a field, jumping against each other, and so on. It is great, right?

But let's step back a little. What is causing this joyful mass hysteria, mere sound waves? Can you imagine if you encountered the scene but everything was muted, and the tens of thousands of herded together sweating and swaying humans were in a field somewhere! Sound hitting the ears of these people, with certain mathematical characteristics — a particular frequency, timing shifts, a particular pitch and combination — vast tracts of it traceable to counting from 1-4 in so many pop or rock concerts.

Take a bird's eye view of this, look at how ridiculous this is — a great bunch of people jumping and swaying to the emitting of previously heard patterns of sound waves. And before you get snobby around the rock and roll concert, the exact same diagnosis applies to a bunch of enthralled, well dressed patrons of your nearest concert hall and their symphonic sounds. The exact same thing!

Why are we moved by sound waves? These are just sound waves hitting your ears, generated by different technologies or instruments (again, all of which have their own evolutionary tale to tell), and we can see with little effort the tracing to our imaginary ancestors again; the coming together of many, the partying or celebration after a successful hunt, the link to sexuality (omnipresent we can assume), the vocal cords and communication, the sounds—some quick like a chase and the heartbeat in a fight-or-flight scenario, some frightening sounds traceable in some impossibly long route to the noise of a beast and why a child might mimic the fear it sees from its parent's reaction to that noise, and how that noise then could be associated with fear—I mean, really, what in the physical universe are frightening sound waves?

Music, in fact, shows us much of our physicality and much of our lack of a special position in the universe.

Should we, then, not enjoy music? Of course not! I enjoyed the concert while thinking of these implications. In fact, I stopped thinking a number of times so I could enjoy the music unimpeded!

We can do this with any human activity—we can appreciate its ordinariness; though, obviously, from a human perspective this can seem really difficult.

We can take comedy, for example, and highlight the communicative function being used in a form of hyperbole, or paradoxically describing the opposite of what it should do, or the licence of being offensive in a comedy setting, and so on. By such means we may discuss comedy, but we don't want to really, for the same reasons as we don't normally want a mathematical representation of music. We could look at art. We could see literature and paintings as two sides of the one explanatory representative coin, as hinted at in cave paintings and the illustrative, inefficient nature of early written language. Of course it is very crude to look at a Van Gogh or a Caravaggio and see this as a mere expression of communication, an attempt to relay a scene, or a reflection on an individual interpretation of a scene from elsewhere; or to see a Shakespearean tragedy as merely at the peak of the rhythmical and yet multi-layered and concise expression of representation of occurrences and their altered sidekicks in fiction. Or we might view the peak of a novel—and see Joyce chip away a little at the veil of consciousness by unleashing the sometimes chaotic script of the interior monologue in contradiction to the more ordered standard external communication—as merely a supremely skilled author reflecting the internally practised speech through which most minds journey each day, and rarely to any real avail. But we do not, we enjoy these things and we abide inside the Trap.

Finally, we can offer our own humorous interlude by trying to trace the fanfare of a major Hollywood production and the Oscar winning performance as an artefact of our early tribal friends, returning from the hunt and conveying representation in para-language and imitation, with the roars of the beast and the screams of the victim, and the learned sadness conveyed at the tribal partner's death, learned from observation and eventually traced back to 'physiological' effects of negative neural-pattern-induced chemical releases and accompanying inertia, lethargy, slouching, sullen expressions, all portrayed by our ancestral friend as he performs the retelling.

A further point on the music concert described earlier is the universality of humans. We all feel practically the same rhythms.

Returning to that Queen concert again though, how did the patrons figure out what song they were listening to so readily? While we must concede the fact that knowing the band (and their back catalogue) might have narrowed it down, how many notes do you normally need on your car radio or elsewhere to identify the song? Where the brain accesses the tens of thousands of tunes it has heard is a bit of a mystery in itself, is it not? Is there a massive CD folder in the brain? Recall the sensation of hearing a song for the first time in years, decades perhaps, and yet you can sing along, the words come to you prior to you hearing them on the radio.

Chapter Six: Ways to See Beyond the Subjectivity Trap 77

How does this work at all? It would seem that the brain, as indicated in the memory-prediction theory, takes note one, and all possibilities of the next note emanate from there; it adds note two and tries to see which are still available; by note three and four their order, pitch, speed, etc. is narrowing down greatly. Your prediction as to what comes next is becoming accurate and satisfying. In other words, you are making predictions as to what will come next and then readjusting if your prediction is incorrect. This is how we see and generally experience things to a great extent, and why so much goes unnoticed (like driving a car for three hours and recalling about five minutes). Because, throughout your prediction (road, next more road, keep hands at this angle and pressure, road, more road), you allow your concentration levels to dip, and so lower levels of certain chemicals and oxygen blood flow is required, and at some level the synaptic strengthening is not triggered significantly… but if a naked person jumped out in front of your car, you would come to attention!

We might also ask what the difference is between music and random sounds. None. We humans recognise patterns, that is all. Again, to a non-human or non-biological universe there is no fundamental difference.

Music and the Memory-Prediction Model can be melded here to liken the brain to one heck of a 'grand' piano. It reacts when the note is pressed, but there are an awful lot of buttons. When the tunes are played the connections between them are recalled. And when the tune begins, the brain looks to complete it, it self-corrects and has inbuilt rewards and reinforcements. It is important to see also that although there is 'a' piano it is not some singular entity that tries to create music—it is a connected system that seems to act as a single entity. Although it may seem exceptionally simple, I believe that this analogy can present a most reasonable centre for a complete theory of how the mind operates, and when coupled with the concepts of the subjectivity trap (with the accidental evolutionary continuous narrative function) we can reasonably look at consciousness as being understood, and also as not being a hindrance to some of the options discussed in Book 3.

Chapter Seven
Dearest

Perhaps two years into attempting to write this the author had an amazing, but not uncommon, life occurrence, the birth of a daughter. And in the five years since I have watched her through her formative years. If I may begin by saying this, I see my daughter as, in fact, a collection of matter. A grandly designed collaboration between atoms of a number of hues and bones and muscles and organs, all encased in skin and hair. Yet, I also see her as my girl whom I love so much more than I would have believed, in my most romantic reverie, possible. She is a constant and unalterable source of unbridled joy for me.

These two things do not sit in opposition. In fact, without the belief in a magic of some sort there is, at least for me, something even more special in this — like it is what we are, fundamentally. I am human and this is me, this is the very pinnacle of me, her awe-filled face at the most mundane of events affects my emotive states more than any shot of heroin, or so-called achievement, or anything else I can imagine.

At some points I have tried to imagine/observe the underlying mechanics of her development. This is a very clear reason why escaping the seeming mysteriousness of mind is so difficult — it is really an amazing magician's trick. In fact, yes you can attempt to imagine the processing that is occurring, particularly when the child is very young. You can begin to imagine the visual relaying and the learning of food sources, recognition of sounds, chemicals being released around the body, components of the organism jostling for energy and driving growth, the weak control of motor movement; improving steadily towards an entity that can run really close to a wall and with no subjective 'I' effort to avoid touching it. I could definitely see the basic instinctual acquisition of language. I mean, how do you *learn* the meaning of words if you don't have words by which to explain the words you are acquiring? And yet it is extraordinarily impressive, to hear children hear a word once and use it in the correct context two weeks later (a feat that baffles your differently constructed mind all those years later) — not to mention young children who can speak five or six languages with the same competence and the same intelligence as their monolingual peers.

And, even if I were to try to concentrate on her and think she is only a functional unit — and she is only a functional unit — it wouldn't make the slightest difference to how I feel, and I would catch her eye and melt! We can understand ourselves without any magic or despair. I can recall watching her grow and trying to imagine the information and its processing. As I said, when she was very young this was somewhat useful, you could see the adjustments being made — you could see as well the rapidly developing brain on an experiential blank slate, but the ability to keep track or observe all of her knowledge acquisition and the myriad processing that is all the while occurring even prior to the development of language is really difficult.

As for language acquisition itself — there are massive amounts of information confirming the functionality of language, but, again, this all happens really fast as it involves all of the human and not just this thinking portion with which my 'I' was trying to monitor. In essence trying to observe this learning is exhausting and something you would only partake in in a couple of minute spells when the child is very young — because you are still using only this slow, clumsy (in many ways) language-based subjective portion to watch the vastly greater developmental stuff occurring in a brain that is increasing in size, capacity, and complexity in front of your eyes.

Chapter Seven: Dearest

This is impressive at the 'conscious' awareness level but, like the world of the big and the small, it is practically beyond our comprehension at the smaller scale of description—down here, we have minuscule and ridiculously sensitive nerves of the auditory system, mirror neurons coercing imitation (something interesting in itself and the reason why you pull your hand back when you see someone else get burned), and dexterous motor controls—we also have organs supporting a digestive system that is demanding from the off (as any parent knows at 3am) and cells that are being replaced continuously. As an aside, in case you feel like clinging on to the humanoid special position in an unsupportable way, there are creatures with better hearing and smell and eyesight and even innate sonar communication, but I am sure your child or loved one is the greatest thing, and there in itself is why this is right, they are *your* greatest thing.

Chapter Eight
Why is this Such a Hard Sell?

I have had discussions with a number of people around the central concepts of the subjectivity trap. In truth, most react very negatively, overly negatively when compared to other conversations on associated topics, even. There are other bizarre reactions that have been encountered, including one gentleman who agreed with all elements of the argument but couldn't bring himself to agree with the argument. This is why the title 'Subjectivity TRAP' is used. This really is something that entraps you. We are thinking about how we think, and using our thinking instruments to assess our thinking instruments. Also we are using core elements of virtual reality engines (what we process and what our conscious voice acknowledges are greatly different) to assess the validity of the virtual reality we are creating with these engines. It is—once escaped—easy to appreciate the depth of the hold of the subjectivity trap.

We ask—'how can I think about thinking? Explain Descartes' I think therefore I am?' The difficulty here is our starting point (or as people famously, seemingly, answer in Ireland when you look for directions, 'well, I wouldn't start from here!'). *The question we must ask is how did we start asking questions.*

But, again, this is extremely difficult. So many philosophers and scientists have stared at this problem for so long. How come they didn't find the answer? Many people posit on a daily basis things like: 'If we can see so many of the functional elements of the brain and body, why do we not have an explanation of consciousness?' Many really have difficulty with this. They wonder how we can be so far away from explaining it. Of course, there is a very much simpler explanation, and that is that our starting premise is all wrong. This is often the case with intractable problems, though this one really is a little doozy, because in essence it requires ceding that the thing that resolves those other intractable problems is itself a problem. *We cannot explain consciousness as we understand it (despite so much ongoing discovery of brain function) because what we assume consciousness is — is wrong, is illusory!*

So, I think the thought, 'I think therefore, I am'. How in good grief can we explain this from a reductionist perspective? By the way, not wishing to antagonise anyone, but I often hear this anti-reductionism and in my ear it sounds just like religious dogma debating evolution. It is a 'look here, it can't be just explained by that' sort of argument. And again, as with the frustrated evolutionist, you are left in a position where you have to disprove that there is a gentle man in the sky above us in an otherwise lifeless universe, which he also must have created but isn't that interested in, aside from setting it up as something to engage our intellect and challenge our faith.

Part of the difficulty lies in a rather obvious point, that is — as the communicative function evolved from gesture-call, paralanguage to early sentence- or phrase-forming and subject and object and so on, part of the initial design was the locus of the 'speaker', and so there was a located 'I', but also there was a benefit of certainty, knowledge, answer-giving, and so on that would require the development of certainty in one's own utterances, and also in their significance. There was an inherent self-centredness in our evolution which would have included self-defence perspectives of the self itself. We are wired to defend the validity of our selves. So, the core of this book is something that is inevitably going to meet great resistance!

It must be conceded this is not an easy task. I am going to ask again to all the readers who do not believe in creationism but do believe in evolution, do you believe that at some point in our past we were simpler creatures than dogs? Think long and hard before you answer.

Those that Disagree

There are certain key opponents to a 'physicalist' explanation of consciousness, and certain key points that are used in these arguments.

John Searle and David Chalmers are often placed to the fore of this grouping. Chalmers is often linked to his so-called 'hard problem of consciousness'. It is very much worth reading all of this, of course. The general idea, however, is that what occurs in a brain is not-computable. There is a vast array of information being processed by the brain, yet it all arrives as one piece. How does it all come together to give this sense of 'I am one person and this is happening to this singular entity'? This is the so-called 'binding problem'.

To be perfectly honest, there is already a massive amount of work offered by others in countering these points. Certainly Daniel Dennett's work in, for example, *Consciousness Explained* looks at a lot of the oft-cited thought experiments around phenomenology. I would suggest looking at Dennett's work; in it he proposes the idea of the multiple-drafts theory of consciousness. This deals with things such as time lapses between what we feel is our conscious 'I' and what science tells us is happening at the neuronal level, by stating that the 'becoming aware' happens repeatedly at different levels. I don't agree with this as a final theory, and feel that it still does not address the locus of the 'I', because by dispersing it—it is still accepting it. I would suggest, though, that many of the 'philosophical' arguments as to why there must be a consciousness, and why it is not to be found in the mechanics of the brain, are dealt with and dismissed superbly in Dennett's work. Again, though, for the majority of commentators we must acknowledge that they still do not believe that consciousness can be found in the brain—I can't write that these days without screaming shortly thereafter, 'WHERE IN THE NAME OF … DO YOU THINK IT IS LOCATED, IN YOUR FRIDGE?!'

I think another very useful read towards understanding the lack of consciousness is the work of Thomas Metzinger. Metzinger, in his phenomenal self-model, explained in *Being No One,* really is shining a light on the illusory nature of what we assume consciousness is.

In Metzinger's talks on the phenomenal self-model — which you can watch online — there is a receding of the smokescreen of conscious subjective experience as we are introduced to a number of neurological disorders and known 'consciousness'-based anomalies, such as phantom limbs and notions of having conscious control of things that everyone else clearly understands as beyond their 'imagined' control.

Metzinger includes explanations of this by showing the core of what we deem to be consciousness as certain phenomena including the sense of oneself, and you being the centre-point from where you 'see'/experience the world. Again, there are some great examples here such as the rubber hand illusion; which hint strongly at the fact that conscious experience itself is a charade of sorts.

The rubber hand illusion works as follows: an experimenter blocks your view of one of your hands with a partition and replaces it with a false rubber hand where your hand would normally lay, and then strokes both the rubber hand and your obscured 'real' hand simultaneously. This causes you over a short period of time to realign your self-interpretation to include this rubber hand as a part of your self — this is in spite of the fact that you know it is rubber; the signal from our eyes and its synchronisation with touch overpower this fact. Normally the reality of the situation is shown in a lighthearted manner when the experimenter hits the rubber hand with a blow of a hammer — the subject for an instant believes that their hand is being struck.

There is a fascinating extension of the rubber hand illusion that Metzinger has highlighted where we sit watching on a monitor an image of us sitting watching the monitor. What is interesting to note here as we move and watch the screen is that our internal representation does not normally include a view of us in this way—we do not include ourselves in our representation of the world.

***

Most people are happy to follow the logic of generally agreed, consensus-held scientific interpretations of the world, but draw the line at allowing a physical explanation of consciousness. But this opinion should be at least open to being challenged.

Humans figured things out. They looked at stuff, they moved it around, and they cut it into smaller pieces. They continued, looking at illnesses and looking at their surroundings; looking at the vastness of space and the minuteness of the world of the small. All the while all they were doing here was looking and trying to figure things out. It is no different than looking at a tree or a rock, and picking the rock up, throwing it, weighing it, and so on. Individuals may decide when they want to opt out of that process of examination and appreciation. If they must opt out of this process, they should use religion or something else like many others—but they shouldn't look down their nose at those that continue the pursuit.

For example, when humanity was looking at the little bits of things, or atoms, some claimed, 'What matter? My head hurts with all of this, it is God's will that things are as they are'. When some countered and reasoned that the Earth was not flat, then others replied, 'I will live my life denying this, and we will see in generations, maybe, if my descendants can begrudgingly accept it'. More recently, when someone suggested evolution, some others would respond, 'You're trying to tell me that I share my ancestry with the zoo's monkeys? Well, you can take a running jump!' Or, 'Perhaps I can in generations go with you on this'. But now, there are people who say things such as, 'When you tell me that the brain, the fulcrum of my nervous system, is the originator of what I say—well, I am struggling a bit too much, I think I may want to call in God, or something inexplicable. However, I can tell you right now, if you then decide my thoughts, my imagination, my planning, happens in my physical entity—well, that is just so absurd, I am going to scoff!'

Perhaps we should add, too, that instead of asking 'how can you explain my thoughts?', you should perhaps try to understand how you could think that in the first place—why would you have such value in your own thoughts? Why would this species invest so much time in defending its sense of consciousness? What is the closest we can get to a non-human-centred approach to this and explain it as we explain every other organism we encounter? Most likely through the prism of evolution. Finally, you should appreciate the importance of communication to our species, its gradual evolution (and our own individual gradual development) and the role of language or communication and synthesis required by the evolved storyteller function and, ultimately, the reasons why this voice would give itself so much authority and would invoke incredulity at any method to reduce its own importance.

YouTube it

It's great that most contributors to these areas (and any others) are giving talks here and there now, to everyone, which are available on websites like YouTube. What you may well find if you watch some of those who believe that consciousness is this special thing—either evidence of the 'soul' and an afterlife, or an emergent non-reducible thing (or evidence of short-sighted aliens)—is that you can sit and watch these voices and all the certainty that they will place in their arrived at positions. You may also look at those who directly oppose them and, perhaps, think to yourself that, for once, rather than have your cognitive communicative function engage with these convincing arguments, you are going to wonder how this is happening at all; you might look at any one of them and ask yourself, 'now, where did you get those words from? I mean I know you didn't have them at two years old'. So there are iterations of the development of language and associations within the individual, but there are also iterations of meme developments across the aeons of tribes, civilisations, and societies.

You will hear people say, with absolute confidence, 'now I will think to raise my hand', and wow, there it goes... it is raised, Q.E.D. I am conscious. I can make decisions. Perhaps aside from this retracing of how the individual framed these words in any event, maybe ask why the individual can't control its own heartbeat, which we know operates through muscle interaction with nerve end plates in precisely the same mechanisms as each other.

It must be understood, then, that when we discuss these things, we are discussing them with our subjective 'I's — with this function that has as part of its operating manual a sense of being in control of the entire human entity, in spite of evidence that we have discussed to the contrary — so we are, in fact, holding the discussion with, and within, the subjective I.

There is a very strong counter-argument from within the trap that there is some controlling element very closely linked to the cognitive communicative function. There is a connection between the 'I' voice and the rest of the brain, obviously, but it is also not nearly as dominant. Indeed, as the colour phi experiment (discussed below) and many other temporal anomalies and abnormalities show, is it as reliable as we think?

In truth, though, if we accept the subjectivity trap and realise the 'less wrong' interpretation it leaves... if we are going to adhere to the interaction of a complex system, traceable without recourse to a designer and emanating from an asymmetric early universe, then we must go to the real trouble of looking at this in the following way. The brain of the individual that has made this most obvious claim has not just 'turned' on. It really has been journeying without pause (bar being in a coma) throughout its existence. Taking the example of raising an arm once again; at one point as a little baby, an individual's hands shot up and down. At another point it saw someone else make this observation, using the language this person had learned, and it included it in its memory as a form that involved the body movement, the word sounds, the previously included 'meanings' and the connection between the two, plus the impact the communication had within the scope of the human 'conversation' (which has been developing from the innate desire for smiles from parents as an infant, or of the desire for touch/affection). And so on. Hopefully you can see the point I am trying to make here, because to fully make it — step by step — the evolutionary path of a meme and its announcer — would take another book I think!

Many of these YouTube talks you may find will begin with seemingly logical assumptions such as, 'I can think about my thought, I can make a decision, I can plan'; these are grounded in our humanity and human experience, which is of course fine as far as it goes… Consciousness is taken as a given and then we must address everything else. The problem is we then proceed to analyse against universal laws of physics and so on.

'We Have No Idea What Consciousness Is'

I continuously read philosophers, neuroscientists, artificial intelligence programmers, and commentators make the above assertion. I believe it to be nonsense, an example of the inescapable egocentrism of the communicative primate, that battles with helio-centrism, deities and designers, purposes and philosophies, and that is enveloped in the 'Subjectivity Trap'.

You should see the vitriol that ensues from any form of physicalist interpretation of the self. Remember by physical we mean that the brain is not different from everything else in the universe in its constituent parts. Or, to put it another way, that this 1.5 kg organ, that is pulsed by blood and affected by caffeine and alcohol and viruses breaking the blood-brain barrier, and that is encased in a skull in a human body on a planet of many other primates and millions of other species, is the one thing that comes from another dimension!

'I can do amazing things with my mind, creative, abstract musings. Explain that from a physicalist point of view — SEE, YOU CAN'T!' This has been shouted about also. In fact, a cognitive scientist once denounced the physicalist view of understanding the brain as follows:

> [T]he computational metaphor of mind needs to be put back firmly into its box by being seen for what it is: a metaphor, and

> not a set of facts about humans. It thus belongs on the shelf with Freud's tri-partite division of mind, and the notion of mind as related to holography, and any of many other metaphors. (This point was made in personal communication. I'm afraid I can't provide a direct reference, but it's not important — it simply serves to demonstrate a popular general view)

I am sure I would have possibly felt this way at some point; but it really, to me, seems inescapable that this is an example of something being dismissed rather than understood.

There are a number of thoughts here that we should probably include in a discussion. As I partly attempted above, reduction from complex thought to synaptic pulse is a very difficult mental journey (pardon the pun), but is wholly doable if a little arduous; on another route one can take a long hard look at evolutionary biology and the biosphere. Then perhaps it would be useful to work from the bottom up from that synaptic potential to the abstract thought. And look at the role of chemistry in life, the core informational repetition and variety of DNA, the origins of our planets, our chemicals, our universe! Then avoid some form of anti-delusionary delusions, take a deep breath, and go for a nice walk.

The Binding Problem

When a theory often makes old arguments sound silly you are probably onto something. So, very many key persons in these arguments will explain that 'the binding problem' is the single biggest obstacle to a 'functional' or 'physical' theory of mind. I think it is worth looking at again. It is the sense that, despite all of these small details that are happening physiologically speaking, and despite the fact that we can see the neuronal correlates of many activities, by what mechanism (physical mechanisms) are these all drawn together to give us this one experience, this sense of 'I' being a collection of these things? This problem exercises so much thought and debate that it is really, really difficult to ever see a solution to it. Of course, this is because there is no binding in the first instance. This sense of oneness is most certainly illusory. This problem disappears when you see this. We as our 'I' or 'communicative function' are not one holistic sense of ourselves, it is just one part of us in some sense caught in an evolutionary accident causing a loop of reflection and it falsely claims totality. Again, this is because it evolved to be a storyteller, and it is easiest to have a singular entity for these purposes.

The Colour Phi

Let's just take one little glimpse at the storyteller revealing itself in the superbly constructed 1970s experiment, by Kolers and von Grünau, known as the colour phi phenomenon. What happens here is that a green dot flashes on a screen followed by, at a slight distance away, a red dot. The persons viewing this see a green dot then a moving dot that is changing colour, and then a red dot. There is no actual motion, there is no changing gradually from one to the other, it just blinks green and then a little later red. This illusion of motion between stationery images, and the blending of the colours, is, of course, how we watch TV. These are not where the real surprises lie though. The really interesting result is that people see the dot change colour from green to red *before* they see the red dot. But if you don't change the colour but instead blink it back on green again, or indeed a different colour, the subjects sees *this* transition. In other words, our conscious experience edits and changes the order of what it sees; it sees red, then green, then feels that it sees green, transition, red, but it learns all this much more slowly than the colours are blinked on the screen, and it comes back with its 'story' of 'I saw the green then it moved and changed from green to red, and then I saw red'. This may freak you out, because it really does show that you cannot trust your 'I'. Your 'I' is a retrospective reporter evolved for communication, that in this case is just blatantly lying to you!

Emerging

Many argue of consciousness that what we have is an 'emergent' property—that is to say, having looked at all aspects of the physical universe and how the atoms that make us are no different to any others, and how we understand the mechanisms behind DNA and much of our biology at this point — and while we can see that we are discovering more and more about the mechanisms of the brain — it just is the case that we cannot possibly be understood by normal methods — there is something special that 'emerges' from the physical stuff. In many ways this standpoint has seen the grounds of logic and evidence move away from underneath it, and one of its last lines of defence seems to me to be this idea of an 'emerging property'.

Emergent

'Consciousness' is seen as something on a grand scale that cannot be explained from its smaller constituent parts. Oh, yes it can; review entropy/order, etc. Please!

This 'emerging' to me sounds just like ghosts or tarot cards to be honest — just clouded in some form of respectability, at least at the moment, for some reason.

What is this emerging property argument? Well, all kidding aside — I don't rightly know!

I think the general idea is that sometimes the outcomes of a system's behaviour emerge but are not evident from the individual behaviour. You may be able to tell I'm struggling a bit here. They tell me that ant behaviour is an emergent property. I say it is not — go and look at evolution. Appreciate how trailing certain chemicals sees certain options for pursuit into shortest routes and forming direction solutions, and so on, look at how birds flying in formation stick to a couple of rules about the gap to your neighbours and following the bird to your left or right or whatever it is.

'The weather is chaotic, and could never be re-modelled'; but again that is not correct. It is extraordinarily difficult because things can vary so much and have so many effects, but given an infinite amount of computation, of course, theoretically you could reverse engineer a day's weather — though why bother? Maybe we need to appreciate that what we are putting down as emerging properties are just forms of saying — 'actually I don't know what is happening here' (or rather how it is happening). But to continue a species-long tradition of sounding certain in an attempt to gain tribal privileges and a desirable partner, role in tribe, and offspring—well, I am going to come up with something: this order emerges from disorder—or, at least, these outcomes are in some way unrelated to their constituent inputs! Is it completely chaotic, entropy? Again may I suggest that you look at entropy and come back and look at emerging properties again—and then stop talking about it. It is a lazy thought.

Getting a Little Angry—If that's OK!

Sometimes I look around and feel angry at the opinions I hear of consciousness. For example, a recent one is something called the 'unified theory'. This argues that the recently supported Higgs particle shows a certain underlying unity to all things, and so (I feel there may be just a slight leap here…) consciousness is what underlies the whole universe, and we are the manifestation of that consciousness. We are one!

I mean, come on!

And what about quantum theories of consciousness? Well, these theories smack of an argument that suggests, 'look consciousness is complex and pretty difficult to explain, so is quantum mechanics, and so they must be related'. Look at the intelligence/order required for a brain again. The people putting forward these theories know the irrelevance of quantum mechanics to large-scale properties' operations. They do, however, in a seemingly warranted manner, say something to the effect that—'how do you explain one overarching awareness that sits atop all of those functions simultaneously?' From a computational point of view this is non-linear. Of course, if you feel that that is what we refer to or understand as consciousness (this omniscient focal point) this is a fair point, but if consciousness is not this at all, not a 'lord of the brain' but merely a part that has claimed, erroneously, to be lord, a continuously running voiceover (a residue of communicative evolution, that continuously prepares to tell its story and tries to wing it on stuff it is not sure of, easily gets confused, is fooled perpetually, and makes predictions and assumptions that are useful for survival, and even for that not always…), then the importation of these desperately non-biological explanations of a biological entity becomes unnecessary, really.

Zombies Everywhere

Another one of the main arguments against a physicalist interpretation of consciousness goes something like this: how would it be possible for us to distinguish between a human and another being which is precisely the same (physiologically, etc.) but is not conscious? It feels pain biologically, but doesn't have the 'sense' of pain, and so on (this creature is referred to as a Philosophical Zombie). I've got to come clean here and say that I understand this argument but it really does give me a pain *and* a sense of a pain in my head — and I can't for the life of me tell the difference!

Now, many raise the flag of subjective confidence at this argument. Merely stating that we could theorise a human who is not conscious means that there has to be consciousness. I would strongly suggest that we are actually, in fact, philosophical zombies. We are the physical creatures we describe. We have their biology and we have a neuronal reporting of pain, as well as a lifetime of stored recall to this concept of pain and to our learned dislike of it. Therein lies a big problem; this is why people believe that consciousness will not be found through artificial intelligence. Many argue that AI can never be conscious. I agree with this sentiment — but, I will add that this is because there is no such thing as 'consciousness', as it is merely our illusory sense of a special position for the interior voice that evolution has bequeathed us, as well as society has schooled us.

This is really the essence of why we are trapped. We do not appreciate our own consciousness for what it is and so our starting point is how to explain something that is indeed inexplicable, but this is because that which we think with this subjective 'I' exaggerates itself to a point beyond explaining.

Our Ancestors and Our Younger Selves Knew Nothing, Really!

We really should look at the ignorance, as we see it now, of much of the history of human thought (stuff like when thunder and lightning themselves were Gods) — and, also, contemplate why it is that we see such a long standing pattern of defiantly and certainly held views over time being dismissed/forgotten. Is it not worth taking a while to pause and question the system itself that keeps coming up with this stuff? To, again, try to step out of being human; is it not much more likely that the human brain is just an evolving biological entity as with all others on Earth, and, more broadly speaking, biological entities are just particular formulations of the elements of the whole universe? Is it not unfair to confer on it the potential for infallibility?

It's Our Trick!

I mentioned a bird's flight earlier. There are new documentaries being aired just now, as of writing, with incredibly advanced and small cameras being attached to birds to follow them in flight as they soar and dive and respond to the tiniest wind current variation. They are awe-inspiring. We believe in evolution for the most part — we believe that the bird's flight, the ant's cooperation, and the fish's gills are all explicable by evolution; just not our communication or thought! This is very hard to justify. The subjective cognitive development function with its executive assumption (and yet, it doesn't even have this — try telling your hair to stop growing!) and its designed egocentricity as it continues to elaborate the usefulness of informing the tribe — this is not evolution?

In the end to really get to grips with what is going on around us, we must try very hard to stop thinking as humans. To try as best as we can to stop navel-gazing and pondering why it is we can ponder, stop wondering if creatures feel pain, are they conscious, do they know who they are — and realise that the reason we are no different from all other creatures and things is not because they are really like us, but because we are really like them. Whether it is an eagle flying skilfully to its prey and controlling individual feathers to aerodynamically pierce through wind changes, or a shark gliding through the air in an endeavour to contort its massive sleek frame to catch a seal, or a human practising its precise control of its vocal cords to relay sounds with learned meanings to many others in a way that elicits emotional responses and contains instructions and recordings of knowledge, the only thing that doesn't fit in all of this is the navel-gazing!

We Already Know the Answer – But Refuse to See it

Why is it that, although, through human knowledge and scientific appreciation of how systems work and fossilised evidence and so on, we can trace species' development and even brain development from very basic to human – and on the other hand through very similar methods we can cross-reference behavioural psychology, neuroanatomy, cognitive science, and the physics of the brain from neurons firing to a vast amount of behaviours and attributes of a vast amount of species and with a vast amount of commonality genetically and in every other measure with humans – yet there remains this unbridgeable gap if you travel these two journeys: the hard problem of consciousness that one species has generated after millennia of communication, interaction, social development, and a predisposition for collective navel-gazing!

Appreciate – Replicate

If we can truly appreciate the implications of the understanding of ourselves as just something else that emerged from evolution, then we may understand that how a neuron spikes when the eye perceives a bright light at a particular angle is clearly mechanical, *and we can see the neuron described as a 'salty bag', with the difference in certain chemical levels, internal to external, creating the electrical possibility of this spike (all being seen by some as possibly reflecting the internal and external sea-based environments of distant ancestors), as clear evidence of functionality but also evolution…* If we could desist from prioritising our subjective narrative as being 'beyond' evolution, beyond another form of the mechanical functionalism that we can see from evolution, then we must see our intelligence as just one particular form of intelligence. For many it seems to be a presumption that every other creature was but a stepping stone to us, that we are the final evolutionary destination, and as we look around for life on other planets (for example with SETI; the Search for Extraterrestrial Intelligence), as I said already, we assume that if the conditions for carbon life exist they must arrive at OUR intelligence.

This will also allow us to appreciate our intelligence as traceable via its physical manifestation. This physicality is something that greater computation and miniaturisation will allow us to learn more about as time progresses. We will also use some of the architecture we find from more detailed appreciations of the brain's workings to assist our own computational devices in terms of energy efficiency and also distributive and parallel processing. Designing computation systems that have an intelligence that is applicable to a range of scenarios is called AGI (artificial general intelligence). It will not need to replicate the entire human entity to serve its most obvious uses. In many of our musings as seen through artistic considerations and, particularly, science fiction, we seem to want to replicate humanoid intelligence inclusive of chemically and socially driven needs, such as sexual and social needs as well as accompanying frustrations.

Now, for the sake of argument let's assume that one can replicate the humanoid form of intelligence (say with a healthy dollop of psychopathic tendencies, or, at least, self-interest prioritisation). One perhaps ends up creating a terrifying creature as seen in these dystopian movies, as it might have these conflict-based human desires and would have vastly superior processing speed. So the question then, I feel, is… why?

Why would we do this at all? Can we not just see technology as technology? That is to say that technology enables us to adapt and manipulate our environment but is non-humanoid. So, we should progress with ever increasingly intelligent and generally intelligent systems, as 'embodied' lately by IBM's Watson, until he (it) is an astoundingly accurate predictor of illnesses and tester of medical interventions, but we don't need to give him the humanoid language-centred pronoun 'I', or indeed include algorithms into this vast computational system with the imitation of the human desire to mate and the psychological repercussions of spurned youthful courting or limerence, or the vast neural networks engaged in bowel movements!

Network of Networks

So to look at the above again, it is not enough to look at the brain as *one thing* to further explain its complexity. It is one organ and in one unique body, but it is not really a decision making device. We might suggest that the notion of parallel processing is a little misleading as it suggests many things work together to perform one function, as opposed to many things performing many or few functions, which is much closer to the human—complete distributed intelligence. The idea of a singular entity lies in the individual nature of the narrator function (as said repeatedly now), and herein lies one of the difficulties in appreciating the actual workings of the brain, as the subjective unitary illusion denies so strongly this possibility. As argued, the brain can only react in pre-programmed ways to sensory input. The programme itself can change from the feedback loop of experience, but we should accept that—for example—although we may be concentrating on a range of tasks, from what we see as conscious concepts (multi-tasking) as well as autonomic activities, via the neural network's control of the body's function or movements and muscle, but we will still drop these thoughts and a lot of these activities if we felt we were in immediate danger. In fact the sensory input would take precedence over some functions that are traditionally seen as autonomic, such as heart rate and even looseness of bowels. There is a hierarchy to which we respond. In fact, highlighting the animalism of us and our history of clothing ourselves like an emperor, why do your bowels loosen when you get a fright? As mentioned previously, it's a residue of the 'fight or flight' mode of your distant ancestors and the ability to run from prey quicker with less to carry! Honestly! This can help dispel our notions of ourselves and our specialness, no?

So, there is a hierarchy, or rather hierarchies, within the neural networks of the functions of the brain—certain reactions override certain others, according to design. Certainly the narrator is an agent in this and can involve itself in some learned forms. Certain combinations have certain impacts, and slightly different permutations will have slightly different feedback. This is where the pretty limitless potential of the individual lies—it is also, of course, designed with evolutionary survival at its core. Again, patterns of behaviour are affected by experience but established by design, giving a sound basis in the importance of; breathing, circulating blood, digesting, etc. but including the random possibility for all kinds of adaption.

The model, then, to replicate this is not one of vast amounts of information being treated with a massive amount of options being observed and disregarded for the right response. It is not, in fact, a single entity at all. This is a network of relatively simple devices—sensory, motor, and inhibitory neurons mostly, connected to other neurons with outputs eventually leading to muscles and nerve endings. This is all encoded from the protein expression of cells from the encoded DNA stemming from the merging of a particular egg and a particular sperm.

Cityscape

Imagine a future town that sits on various tiers with a complex infrastructure that has evolved over its centuries of existence, and particularly over its recent technology-based 'make-overs'. Part of it rises over rivers while others go underground. The town has been designed and redesigned in recent decades with sensory technologies and communication technologies and networks of various sizes which participate in different scenarios.

Initially these included traffic signals that could assess traffic flow and change accordingly; then further technologies were added, the whole town (or all of its walls, pavements, etc.) were given interactive, overlaid real-time maps; there was an irrigation system that underlay all of the houses and other individual properties and so when a fire was sensed (by temperature analysis systems) water was immediately rerouted and released.

Various components of the city could, via networks, 'speak to each other' or were networked to act in synchronicity in certain circumstances and to not do so in others. As time went by technologists constantly reviewed what they termed the 'intelligence' of the city. They further programmed networks so that there would be a central cloud-based executive decision making system that would instantly conduct internal reviews of overall system efficiencies and analyse, over time, preferred courses of action in given circumstances. The executive decision making algorithms had access to 'communicate' with all other dispersed and interconnected technologies that had been put on the 'grid' of technologies that were now in this city. Sometimes the executive decision making programme would be subservient to the smaller operating networks (a good example of this was if there was a fire—the building involved would override whatever clever data analysis or decision making the executive programme was working on. Of course, this was 'understood' by the original design of the overall city as it was appreciated that this was a city made of fibre optics and timber after all, so fires must be dealt with as a matter of urgency!).

The city was sometimes the recipient of foreign bodies that flew in on the wind and threatened to damage the overall system. The city 'learnt' how to deal with most of these objects by trial and error of its inbuilt alarm and defence systems, and became good in most cases at deflecting their course or stopping their threat. The city had a whole series of alarms and signals which could be relayed within the city itself for mutual understanding of traffic or air pollution or any dangerous nearby items.

Over time other cities with similar technologies emerged in the vicinity of this city. Each individual city system began by using its sensory technologies, including cameras situated all around the city and all around its boundaries, to 'learn', as outlined above through analysing its neighbours' impact on its own circumstances. For example, its monitoring and reviewing of its own fire alarms detected a pattern when it discovered that a fire in a neighbouring city led to a fire in itself as the fire could spread from one city to another. And so its self-correcting, protective algorithms began to use the neighbouring city's alarm system as its own early warning mechanism. It saw that the other cities were using similar systems and technologies. These groups of cities began to communicate and co-operate into one larger network. Over the aeons after this the technology increased in complexity — much of it was designed around not only monitoring itself and making decisions within its own bounds but developing a repertoire of symbols to and from its increasing number of technology-supported city neighbours.

OK, we get the cell then neural workings explanation here via an extended analogy, but what about the initial intelligence? How could inanimate things become even simple cells, and why do we have these things, cells, or atoms, or a universe at all?

Remember the journey of Book 1: from symmetries breaking, the big bang and inflation science can account for events as they unfolded back to tiny fractions of the first portion of a second. The journey from elements born in the inferno of stars and the chemical bonding for RNA and eventually DNA also seems, to many, to be explicable now. It does seem unlikely and the vastness of time and chance over probabilities must be included to our sensibilities, but just remember that we often bring human sensibilities and philosophies to the wrong places. It is often claimed vehemently that there is a fraction of a second that can't be explained, or that there is a moment of spark into 'life' that can't be pinpointed; and, indeed, that the fuzzy nature at the scale of the quantum means that all must remain forever mysterious. However, it feels that if we were taken by the hand into a lab and given a magnitude by magnitude sense of scale, we would quickly enough discount the philosophical manifestations of some of these debates, and understand them to be ill-placed human contemplations. 'Why are we here?' finds itself in all manner of considerations; and by itself must be understood as an ego-centric component of the Subjectivity Trap.

Everything is Computable.
Oh No It's Not! Oh Yes It Is. Oh…

If you are in the groove at this point you will have no problem at all in appreciating that not only can we say that everything is computable, but that it is ridiculous to not say so. 'Wait a minute,' a bit of you may linger '…what of love—you can't compute love'. But, you are computing it! Or rather it is being computed within the entity of 'you'. You, an individual made of atoms using neural networks and a synthesis of chemical- and neural-based 'emotions', and drawing from a plasticity-based memory, a social inculcation and education, and the evolution of the communicative function's ability to predict and imagine, you are computed all the time from moment to moment. Computed just means occurring—when you understand the informational base of matter and the underlying layer upon layer of physics that lead to you stating 'love' or, indeed, 'you can't compute love'; ironically, you must conclude otherwise.

An End of THOUGHT!

Perhaps one of the most profound concepts of comprehending the subjectivity trap in its entirety is that, once we begin to think, we are by definition within the trap. Of course, this is certainly not a means to suggest that thinking should stop—doing this for just an instant is something we seem, in any event, to have massive difficulty achieving. We spend many hours in meditation, religious contemplation, listening to music, getting massages, and so on; absorbing sensory stimuli, externally and internally originated, trying to lower this conscious narrator—and in many cases end up feeling refreshed or blissful (transcendent!) when this is achieved. But leaving that clear indicator of the normality of thought, let's look at what we think *with*. We think with our thoughts. If we follow the logic of the world as we have discovered it, we must arrive back at the origin of this process and no longer assume that what we think with is special, but accept that our thoughts are a part of everything, and the whole idea of priorities and a unique position being given to our thoughts comes from within this thinking and is erroneous. If the thinking is just a function of our evolution and interactions (as we must admit) then once we begin thinking we have entered into this wonderful and hugely useful, but imperfect, activity. So, in summary, we should seek not an end of thought, but perhaps an end to its uncontested supremacy.

The Importance of These Thoughts

We hold our conscious voice very dear to us, undoubtedly. But, and I am sure this will not exactly please the reader (yet again!), really think about your thoughts. Think about the last day or week or month. What have you really thought about? How much of it was repetitive? How much of it was non-superficial and non-trivial relationship-based stuff? You thought about a sport, or an artistic endeavour, or something historic that had been brought back to you — OK, so you are not a slave to the anaesthetic of reality TV — you are an 'intellectual'; again, though, the same question. How much of it was relevant to anything? How different would the world be if you hadn't been thinking that thought?

In the impressive scope of human development in areas including transport, communication, and manipulating our environments through tools and technologies — what percentage of total thought really was directly involved — what percentage of human thought has demonstrably added to the sum of human knowledge, has improved the human condition? Somewhere (and admittedly dubiously) someone estimated humans have 12–50,000 thoughts per day! Let's just say it is only one thousand. Overall this is 7 trillion human thoughts per day — perhaps up to 350 trillion per day. Anyway, I am not going to stretch that over the entirety of human history but, as an exercise intended just to question the supremacy of our thinking, hopefully it makes a useful point. Perhaps we can couple it with the idea that individuals will always change their minds on certain issues over the course of a lifetime. Yet, in the vast majority of cases, they are certain of their position at each point. It should be clear that we place too much significance on our own thoughts — perhaps we feel that they are all 'we' have. But let's examine how we might be erring in our broad perception.

Just in case you are feeling like you agree with the general point, but that in your particular case the thoughts happen to be extremely important, let's try another argument. Maybe you're a political leader, but at a certain *scale* the question still remains; there have been tragedies that have affected the world's population and a couple of meteor impacts that manifestly changed the history of this planet — how often have individuals, percentage wise, fundamentally shaped the course of the universe or even the Earth? So are your thoughts really so important? If someone could record and relay your thoughts of the last week and you were to delete as much as you could, particularly irrelevant or repetitive material, how much time could you condense the week's thoughts into? Isn't this further evidence that what we are talking about here is a residue or a by-product of the increasing evolution and complexity of the usefulness of our vocal cords?

Fit Philosophy

Where does philosophy fit from beyond the subjectivity trap? What is it all about? How did we get here? Without being blithe but to a great extent — it does not fit.

The question of why we are here must be seen in the evolutionary context again. This is an expression by a communicating being. An exercise of the muscles around the vocal cords with clearly many associations, both for the individual and also socially. Again, this is very difficult to even begin to trace. One could, however, beg the liberty to suggest certain assumptions from the evolutionary drivers, such as that *place,* would emerge from the gesture-call creature and the physicality of highlighting (for example) the position of prey. I am here — I was there — seems easily central to any early story-telling.

So the contemplation and complexity further evolves, and millenia upon millenia later we have philosophers such as Socrates impressing and developing such topics of our role in the society into which he grew. And on and on we go — to an individual around the age of puberty posing higher-end abstracted questions they have heard previously, 'why am I here?'

But certainly, as will be suggested in Book 3, the question should evolve away from the obsessive 'Why are we here?' towards 'What are we going to do, being here?'

Tiny Philosophy

As discussed, there is a tendency to infuse conversations about what lies behind the smallest particle, or force carrying particle, or particle generating force, or the underlying field of a Higgs or whatever; to feel that we must continue downwards to find meaning. As I stated, the meaning to my mind remains 'information', or that what must remain is information to allow for diversity; but we do have this difficulty with nothingness, or something beginning from nothing (though physics of today seems to have an answer around negative energy even for this) —really when we speculate downwards like this, I ask again; are we conflating environmental understanding with varying philosophical arguments?

It would be useful for people to again travel the journey from something small, down to something smaller, down to something beneath the naked eye, and then massively reducing in size again and again until we try to imagine what is left. Should we not ask, why is what happens on these scales really able to make us consider large-scale 'creators' and so on? Would an honest reflection be, 'I have no idea but when it gets that small I am pretty sure that we have to doubt it has philosophical implications, that the understanding of our complex structure occurs up the magnitudes, and that the difficulties of causation and nothingness reside, more than likely, in our limited perception'?

This brings me back to another difficulty from within the subjectivity trap: according to our subjectivity trap, the brain is the pinnacle of evolution, and inexplicable by normal physics, and it is also a perfect instrument for comprehension. So therefore we 'believe' there is nothing that we can't understand eventually or that we don't have some kind of divine right to understand, ultimately. It is unpleasant territory again, but how would something that comes into being through evolution be anything other than just very useful? Why should it be perfect as an instrument of universal discovery? And so we must accept not just the subjective lying to us about our own individual and species significance, but we must qualify everything we know as best known and agreed upon by human minds. It may well be uncomfortable, but it is logical (or less wrong). To disagree with this you must state that the brain is a perfect measuring/observing instrument, and why!

The Impact of Brain Imaging

Often times there are headlines in the press nowadays, which lay claim to precision in brain imaging that scientists can only dream of at this point. This is still mild compared with the nonsense that passes for reporting of genetic 'breakthroughs'; science 'discovers a gene for creativity' type of thing! Much of this book may be described as attempting to demystify the human circumstance, but it is not at all suggesting anything other than the complexity and interrelatedness of these entities, and so a gene for something so prone to cultural effects, aside from the gene having so many levels of involvement and other genetic causations not being considered, all lead to these reports as mostly being exercises in wishful thinking.

The resolution of current in vivo imaging is still much removed from the cellular level: though there is an inevitable path being worn in that direction, it is happening in line with exponential technological capacity growth, and there are extremely talented people doing simply amazing work — people that we pay just a fraction of what we pay a business CEO type, bizarrely. Anyway, I digress. What can be seen using imaging now is a pattern of activity; so certain 'areas' of the brain are active during certain tasks or thoughts. The 'areas' in traditional fMRI scan resolutions are millions of cells, and so billions of synaptic connections, but still, by looking at the pattern as closely as possible, and seeing the same activity represent the same pattern, then immediately the functionality should be apparent. In any event, the volume of work in brain science is increasing really rapidly, and has just (as of writing) received a major boost with similar competitive bids across the Atlantic to map the human brain. Believe it or not, a very large number of experts see this as a way of understanding the *brain*, and not the *mind*, to this day! The preciousness of the subjectivity trap ensures that we need to create this concept of mind and extract it out of the physical world... even for some scientists of the brain!

There is so much data coming from all over the world now on neuronal responses and experiments that it is literally impossible to keep pace; thankfully and necessarily, again, our subjective 'I' will look for summaries and shortcuts. I would like if I can just to take one interview with one such neuroscientist on one particular experiment to highlight some of what can be seen to be a trend of appreciation of the role of the neural network in building up 'our' world model or our Perception: I feel all of this is valid and so have just included the extract in full:

> Diamond and colleagues monitoring rat neuronal behaviour could see when a rat mistook an object for another object how the object of the error was encoded exactly the same in the neural firing. Diamond goes on to explain; 'Our method is so accurate that when the rat would mistake one object for another, the decoding would also indicate a different object from the one actually touched. And this happened because the representation made by the brain—and, as a consequence, our decoding—appeared like that of a different object. Hence the error.'
>
> Diamond's team has no intention of stopping here. 'In real life, we generally recognize objects using more senses all together, in an integrated manner. We use touch and sight at the same time, for instance,' explains Diamond. 'For this reason we are now working on new experiments employing more neurons, with more complicated stimuli, and more senses, to build "multimodal" representations of objects.'

This kind of 'mind reading' carried out on rats' brain by Diamond and his colleagues is important to understand how the brain forms a representation of the world. 'Each one of us perceives a physical world outside ourselves, yet actually all we have at our disposal to create an experience of the world is the representation that our brain makes of it through the input of sensory organs,' says Diamond.

To understand that such a representation is at the very least partial it is enough to think of all the information about the world that escapes us all the time: for instance, we are blind to infrared and ultraviolet rays, we are unable to hear certain sound frequencies or smell some chemical substances or others. Some details pertaining to the physical world are completely invisible or, to put it better, imperceptible (others are interpreted incorrectly, like visual illusions, for example).

This is a further demonstration that what we perceive is not the physical world in itself, but the neuronal activation the world evokes inside our brain. (Diamond *et al.*, *Journal of Neuroscience*, April 2013)

Connectomics Is Where It's At!

SRGAP2C—this gene is uniquely replicated in humans, all other mammals having the one copy—is linked to cortical development. This is a gene in chromosome 1, which apes have one of, and we have four of. I happened across a *Horizon* documentary called 'What Makes Us Human' and there were a couple of fascinating insights in its presentation.

Firstly, the non-controversial polite presenter claimed that everything that makes us human is contained in the connections of the brain, and she stated this with absolutely no qualms, no dramatics. I, for one, was bowled over by this presentation of emergence in mainstream commentary.

Now, there was, of course, a whole portion about how amazing this structure is, with trillions of connections between neurons, etc. It's interesting to note, though, how perceptions change; if this lady was a fringe speaker at a fringe festival just a decade or two ago she may well have been greeted by derision for the extremity of her claim that neural connections hold the key to everything we experience.

It is interesting how this journey from absurd to acceptable normally progresses. When something radical is suggested as coming through scientific and technological progress it is derided as ridiculous and impossible in the years leading up to its achievement, but then to keep our own sense of 'being in control' it is very quickly derided as being predictable, nothing new, and in some way not that good almost immediately after it is released/achieved, etc.

What struck me as even more astounding from this presenter's commentary, though, was a little throwaway demonstration of how engineering this gene (SPGAP2C) into a mouse neuron caused the neurons connectivity to increase!

But this concept, that the connections of the brain are what 'human-ness' is all about—brain, mind, etc.—is known as connectomics, and it is growing in significance, and funding.

Interests in the Brain

I might just mention that there have been, of course, post-mortem examinations of the brain that have been carried out for centuries. Two major issues have always been tissue damage and available technologies for high resolution. Kenneth Hayworth (mentioned previously) and his team in Harvard, for example, chemically preserve the brain and then carry out a 'slice and dice' method that you may wish to read about, as with the other brain imaging techniques that will be more and more publicly consumed over the coming years (due to the EU sponsoring Henry Markham to the tune of €1 billion to map the human connectome, and President Obama assigning €500 million to their brain mapping attempts). Essentially, though, the volume of neurons in miniscule space is mind boggling, really; so the fact that Hayworth can present a computer generated 3D map of even small portions of the brain is pretty amazing, but they can, and again you may want to look into this work.

To Conclude

The problem of a credible theory of mind is discussed in philosophy, neuroscience, AI, and very many other fields, and it is largely held that there isn't one.

I am stating boldly that this is it. It is not easy or straightforward to articulate, and is not a particularly uplifting one, but I will bet my own sanity on it being correct!

- Consciousness is not a real thing, though it is important within its own generated reality, and has real implications.

- We are entrapped by the generation of self-importance in one function of our brain, that cannot disown the actuality of its own ordinariness.

- We are not justifiably better or substantively different, in a non-human or universal sense, than another animal or, indeed, any other matter we know of — in fact there isn't even a criteria for comparison, other than our naturally prejudiced one.

- We are a collection of matter, like plants that move and interact as a result of their sometimes shared and sometimes more unique evolutionary story.

- Much of our brain's functionality has very little to do with the portion we attribute the concept of 'mind' to. This portion is sometimes a contributor to executive (or coordinated) decisions, but is more often a 'slightly late to the party' reporter, rather than the holistic proprietor it claims to be.

- Speech is an evolutionarily refined species attribute, and not in any substantive non-human-biased manner different from flight, procreation, sonar, photosynthesis, and all the other living entities' evolutionarily designed functionalities.

- Communication was a key benefit to our species and became heavily rewarded in our communities, and rehearsing the speech evolved into thought.

- Our thoughts or 'communicative function loops' and 'speech rehearsal' are useful to us, but are one part of what we are and are in no way more special, universally, than anything else.

- Nothing 'matters' in a universal sense, other than what we say 'matters'.

- We have a mental representation, or neural network correlations, to our surrounds that, naturally, takes us as its central point but does not consider its own internal operations.

- We have a history of centring ourselves with beliefs and concepts such as helio-centrism.

- We, erroneously, place our communicative cognitive function at our centre — executive leader of the brain, in fact — brain's entirety, single most important thing in the universe, even to the point where it claims occasionally that if it didn't exist then everything else (despite the absurdity of this claim) might also not exist.

- No one is watching us, aside from us. The odds of another journeying and communicative being in the observable universe are minuscule.

- There are no practicable limits to what we can as a species do (either good or bad). In fact, these are concepts that abide in us. In truth we can use the lack of an external observer to justify the worst acts or we can appreciate our shared Earth, our history of development through cooperation, and the logic of only striving to make things better and better for us all.

- We are at a time where we can see more of the world around us, but, more importantly, we are seeing much more of the world within us. Eventually the mirror of our ordinariness and, also, from our perspectives, limitlessness, are becoming clearer.

Interlude

What is Technology Anyway?

Here we digress significantly—the story as writ continues in Book 3 if you wish to skip ahead.

Technology

Continuing on our examination of what we mean by human intelligence, we now consider what artificial intelligence is and inspect the meaning of rapid development of technology for the human condition in ways that will be considered again in Book 3, 'Beyond the Subjectivity Trap'.

Computation

When discussing the historical developments of twentieth-century computation or the notions of artificial intelligence and the centrality of information, I believe it is essential to appreciate the physicality of what is occurring. There can be a tendency to call this 'Virtual' and then distance ourselves from the mechanics by implying almost 'Alien' as a synonym for 'Virtual'.

Cyberspace, virtual realities, the internet all seem to give the impression that information technologies, communication technologies, and the computer revolution are all based on something, as with consciousness (though for different reasons), 'not physical'. So, it is essential to at least accept that what we are dealing with here is physical, but the speeds and tininess of what we are talking about (as with the complexity of the brain

Interlude: What is Technology Anyway? 113

and the magnitude of the solar system) are more awesome than most issues we concern ourselves with.

To begin with, and I can only offer the layman's guide, again, let's look at how our computers operate on the binary system. As I've pretty much stated already, we may express everything in the universe as information. At atomic and subatomic level the existing differences that combine to give us the variety we enjoy in our environment are formed. They can be enumerated simply, by *some of this with some of that and a little of the other, and so on*.

If you would, consider how a computer works at its most basic level and consider its basic code, or binary code, which uses just two numbers, 1 and 0. Now imagine 0 and 1 as an on/off switch for a light. So if you are to hit 1 the light is forced to come on and with 0 it is forced to turn off. The connection from the switch to the electrical output and the ignition of the light source is in situ from the original design. Simple enough? Well, that's it really. That is basically all there is to understanding artificial intelligence!

So, early on our calculators might be programmed so that, just as 1 means on, hitting 1 and 1 will cause a series of stepped processes which result in the screen of the calculator showing 2 (the appropriate pixels illuminate in a different colour to all others to give the lit shape of a 2). And if on pressing 2 the screen immediately allowed the internal computation device to recognise this number again as being devised of those earlier component parts then we could by adding a 1 have the computer understand (by 'understand' we mean it is pre-programmed to do so) or follow its programming to present us with the visual representation of 3.

Eventually, by informing the calculation system of all mathematical shortcuts it might need to use our mathematical knowledge, we have a calculator. On our computer, for instance, one instruction is to open a file, one is to increase the volume, and another is to play a bit of music. Indeed, each note is expressed somewhere in that whole matrix of commands as a particular set of 0's and 1's which informs the computer to express through its speakers that particular note. Now, this is a whole bunch of 'ons' and 'offs' and pre-programmed reactions and yet we could mirror all of this without electricity!

Consider a device that has a left pedal and a right pedal. Both pedals are connected by a pulley system to two buckets. One bucket has a bunch of gold while the other has some poisonous toxin. If you press one pedal you empty one of the buckets. If we see this as a game show scenario or something like that we would have no problem understanding that what we do 'informs' the buckets by exerting a force on them to 'do' something. The string attaching the buckets and pedals is the 'designed' component. We would also have little trouble appreciating that we would love to press the correct pedal. For the heck of it let's say that if we press both buttons together at the same time we open an escape route and we get the gold and win a holiday. Again, we would like for someone to tell us this and we can see that it is a new and vital piece of information pre-programmed into the system.

Just with the two numbers we have a meaning for hitting 1, 0 in that order or alternatively 0, 1. But the code can be written much longer than this. Clearly then, with the many permutations from just a left and right pedal or a 1 or 0 we can see that e.g. 1,1,0,0,1,1,0 could mean something else entirely, and so on and so on *ad infinitum*. We can see this pulley/pedal system being linked into the inner workings of a PC leading to images on our screen of increasing complexity and variety via appropriate illumination of the innumerable available combinations of pixels. But, take a movie on my computer… how does it play?

Well, from this same vast set of 1's and 0's that inform a pre-designed array of colours to hit the very large numbers of pixels on your screen to illuminate a particular picture. Each pixel can be illuminated in a whole array of colours that in total present a particular picture on the screen. The pictures are discrete and are followed quickly by the next picture — of course, it happens so quickly that the illusion of movement prevails — the moving picture (as a curiosity this is how your eyes function also, your brain learns and interprets movement).This all stems from the rather laborious binary code, where we can see '1' as simply replacing 'turn the light switch on' and '0' meaning 'turn it off'.

Simultaneously a certain set of sound waves hit the speakers on your computer and you sit and enjoy your movie. To illustrate this someday perhaps someone will design a set of levers attached by cable to tiny light switches and speakers, and someone will, by pulling one lever after another after another and in a particular in order, play the movie! It might take a while, in fact many, many years; and would certainly need to be sped up significantly before it would make sense to a viewer!

While we *could* play the whole video by hitting pedals it would, as mentioned, obviously take a ridiculously long time, and so we harness the immense speed of electricity and we build order and order on top of yet more until eventually by moving my mouse over the play button I can programme the playing of the whole film with a click. What I mean here is important if, for example, there is a particular image that I want to show on the screen and if I have shown it already, then I should employ a copying of all the code function, a shortcut. Perhaps I have two images coming on screen where one is slightly different than the other. Rather than build the second image from scratch then wouldn't it be easier to copy it and add in the differentials only? At higher levels these codes can represent nodes similar to the brain's workings, and so with one simple instruction the underlying programmes each are enacted with their individual pre-programmed complexities. Of course, I may also use my computational device to just tap into much larger data stores online or perhaps in the vastness of the cloud. Again reducing redundancy and increasing efficiency. And I access my YouTube, for example, press play, and view and listen to the recording, practically instantaneously. So not only can the information or instructions be performed at astounding speeds, but, of course, we don't need to write the entire code again each time someone watches the movie.

We can also look back to Book 1 and see the form of the very base elements of our current understanding of the physical realm, be it atoms or quarks—and looking at how these atoms come together to form all of our elements, and how pressure and heat and magnetic and electric and possibly other fields all combine and interact, we can see that, again, there is a very basic information pattern at play.

We can see that an atom either has one electron (i.e. hydrogen) or more (other elements), for example. From these elements we get trees and storms and houses and mountains, and anything you care to add. While clearly describing the world or the universe requires an incredible amount of information, it is still the same computable process. Indeed, perhaps most complex of all is the inner workings of us, living creatures, humans. So, if one does not accept that we are incomprehensible because we have an essence that is not atomic or related to this universe, e.g. a soul, etc., then we must accept that in principle it is some of this and more of that interacting with lots of that while retaining a large percentage of continuity of pattern, just like anything else.

Let's take another look at the notions of intelligence and artificial intelligence. As we outlined previously, when we hit the light switch the light comes on because of the initial design from the manufacturers and the electrician. If, however, I open a door to outside in a darkened room, that room is also illuminated—we would call this natural light. The photons have the same basis as the light switch model; the photons are being emitted from nuclear reactions in the sun in this instance. The only difference in reality between both originators of photons and illumination is the involvement of humans. In a very prejudicial twist we ascribe one as intelligent and the other as natural.

In fact, a human, an ape, an amoeba are all intelligent creatures, but so is a star or a rock. The truth is that intelligence is the predictable or continuous ordering of matter. This is the world we are familiar with and that we are of, the world of things. We can say then that 'everything' is in some way intelligent; however, there is a caveat to all of that—Persisting Entropy.

In fact, all of these things are counterbalanced by many magnitudes greater of 'non-things' where the necessary structure for our recognition does not exist. This brings us back to the principles of entropy, the underpinnings of quantum uncertainty, and probability. In fact, as with the falling glass example in a previous chapter, the numerical tendency is towards disorder, and theories such as the 'many worlds' theory or the 'sum of all histories' where people imagine that there are an infinity of universes that never were and that are completely different and unreachable, suggest this.

In essence the point here is that everything we know is by definition of the same essential form as ourselves, but all indicators of logic suggest that this stuff we know is not and never would be 'everything'. In summary, then, all that we know can never be all that there is to know, and as being seems to be possible in forms definitely hidden from us, we know that we can never know everything. Also, while we place massive importance on intelligence, and claim it as our own and, with descending levels, to a few other species, we are again guilty of homo-centrism and should replace 'intelligence' with 'order' and 'continuous order', or patterns, and manipulation, interaction, and so on.

I hear a lot also about us being the only species that can manipulate its environment and learn—both false claims as a cursory glance at many other creatures will show—but we are, relatively speaking, *more capable* of these things, and at greater scales of complexity, certainly.

What is Technology?

When the word technology appears before our eyes there is a reasonable chance that we imagine circuits and plugs, computers and mobiles. It is important to begin by trying to get the fuller meaning of the word technology.

When our distant ancestors used flint to shape spears these were technologies, these were materials outside of their bodies they used to achieve goals. Technology is the means by which our species improves its lot. Hunting, transport, agriculture, and industrialisation are not good for their innate aesthetic—but because they have made our lives more secure and satisfied (this comment acknowledges, of course, the need to include harmful technologies; but, of course, this is the will or intent of the particular humans). We can see that they enable the major transitions of our species. Spears led to our dominance as hunter gatherers (from our first *Homo sapiens* ancestors to just 10,000 years ago), ploughs and other technologies led to our becoming non-nomadic farmers, and the industrial revolution moved us in significant numbers to being urban dwellers, mass producers, and ultimately mass consumers.

What is AI (Artificial Intelligence)?

Within the technology of the industrial revolution we saw clear examples of artificial intelligence (which, to repeat, can be erroneously seen as some intangible computer-based notion). If I design a machine in terms of levers and gears, cutting instruments and pressures then, by placing the component parts in this particular arrangement, each time I send a material through the machine an altered usable item appears at the other side. How is this artificial intelligence? Well, by design. Anything that we design to exhibit this type of behaviour is intelligent—it does something useful, creates order, and we call it artificial because it is designed by man and not resulting from evolution, like, say, a tree, which is a very much more complicated manipulation of matter.

A Computer and an Oven

Now, a real game changer in all of this was, through greater scientific understanding, being able to harness not just the steam and coal power of the industrial machines but also, like Benjamin Franklin's kite, the awesome speed of electricity. As just noted, computers are physical entities as much as you or I. Indeed, coincidently, the first computers were humans; and this is where the name was derived; from people who added and subtracted lists of numbers, or computed!

We just designed the artificial 'computers' to do these additions and subtractions for us at greatly enhanced speeds. And we progressed to allow computers to perform amazing tasks such as reading and understanding and using vast quantities of data, where a computer like Watson can read an estimation of the entirety of human knowledge in a matter of days! And where processing units (as of writing) hit the measly mark of 17.59 Petaflop/s (quadrillions of calculations per second) – wait a minute, what does that mean? Well, remember 1 means on and 0 means off, and combine this plus this (these are operations, simple switches). Well, this computer, which is formed by combining a bunch of computers, is able to perform 19,000,000,000,000,000 of these per second. When looking at a number like this it really doesn't make any sense does it? But, like our big space at the start of this book, let's force ourselves to face the reality. Remember, as we move from 219 million to 220 million that is one whole million operations. But, bear in mind that when we move from 18–19 *billion* that is not one million more or even two million more but ONE THOUSAND million more operations. Right, well, one trillion is one thousand billion, and as we 'get off the pot' try to absorb that one quadrillion is one thousand times those nonsensical numbers. Just nonsense really isn't it? Stuff moves around at significant fractions of the speed of light!

What is in a Word: Exponential

I cannot tell you the number of times I have heard politicians say that something was growing exponentially (mostly debt), but in all cases they are brutally wrong; these things are growing an awful lot (which is what they wish to say), but they are not growing exponentially. I heard a nice explanation of this: a guy asked an audience, 'if I took 31 steps where would I be?' And the audience guessed 31 yards away. 'OK,' he says, 'if I take one step, then 2, then 4, then 8, and so on, if I take 31 exponential steps with an exponent of 2, where would I be?' Maybe you would like to take a guess… his answer? To the moon or around the world eight times!

In 1965 George Moore—co-founder of Intel—observed that the number of components on an integrated transistor had been doubling every two years since 1958. He boldly predicted it might continue for another ten years (it should be noted that what a computer can do, the number of computations and so on, is pretty much directly related to this). The key thing to do now is recall what we just said about going to the moon and what have you with the word 'exponential'. Well, Moore's prediction is still holding true now! Hence the absurdity of what we can do—it is really a question of finding applications suitable to the raw intelligence horsepower we already have.

Remember though, all of this is designed, artificial intelligence, just as is an oven where energy is channelled specifically to heat bread. Remember there is also no difference between natural and artificial. This is just misunderstood. If we blow a leaf or the wind blows it, there is no difference.

I hope that what you can take from this is that things occur —they are computed or we can compute what happens to them. Forces and interaction and complexity and so on. Keep in mind that there is great difficulty in many things, but impossibility is very different.

What is of most interest for us is the dual ongoing pathways of (a) bettering our circumstances, food, and clothing and shelter and transport and entertainment, and medicine (and the understanding of ourselves, our biology, our central nervous system that that requires), and (b) the aforementioned crazy advancements of technological control and the massive help that this vast ability to research and simulate and trial and underpin macro-level infrastructures and so on can do... Speaking of which, just as of writing in 2013 I saw *an algorithm, an artificial intelligence, had devised and tested a scientific theory all by itself! It was a robot called Adam at Columbia University and it happened back in 2009!*

There is a theory espoused by Ray Kurzweil and others back in the 80s, and adhered to by many now, called the singularity, where computers outperform human intellect, and in going beyond it dream up worlds we cannot possibly fathom. It is jokingly referred to by many within and without as the Rapture for the Geeks. On the negative side of this 'big' idea, it does seem to have imbued some with a sense that this is all destined and does not now need work. In the positive side, it has definitely brought to the fore the immense changes that are possible through computation's rapid (mind-blowingly rapid) expansion. There are undoubtedly possible dystopias and existential risks at play here, and we do seem to be adhering much too closely to our science fiction violence-packed creations as templates for the future we are 'programming', but again there is also a little bit of stultifying human centrism in this theory also. But it is amazing — it is just simply amazing what is happening in and around electronic technologies.

It seems fairly, and typically, self-centred to assume that we are some sort of aspiration for any processors we devise, though surely we suspect that — with our propensity for poor communication, acting with minimal information, group mutual annihilation, unproductive systems generation, et cetera — it would be a whole lot better if we could devise, for example, an intelligence system that has the capacity to hypothesise, research, and test cures to remaining diseases in a strong scientific manner and can do this within a hyper-time-processing context without needing to mimic, so thoroughly, the humanoid processor. Why would we want a computer of any form to be insecure, to confuse sexual desires with violence, or to need the toilet or sleep?

To be fair to the considerations of the dystopian potential of such an outcome; of course all of this does not necessarily eliminate the likelihood of persons pursuing the creation of process generators that can act with the pre-programmed intent of causing harm, and perhaps learning/adapting to refine the process of causing harm. Einstein's warning that 'our technology has far outstripped our humanity' is again very relevant; we can take humanity here (in the context of no more reliance on a 'special role' for humans) to mean our ability to cooperate to our mutual benefit; to be pro-human.

Book Three

Chapter Nine

Beyond the Subjectivity Trap

So what if there is a subjectivity trap, what difference does it make if our self is just a component of our biology and in no way mysterious or requiring a soul, a Cartesian duality, or a 'rescuing' from quantum mechanics?

I believe a lot of our inertia comes from these ideas that, well, we can't explain consciousness, we can never emulate or comprehend the brain, we are mysterious and doomed and should offer ourselves to a hope of an afterlife. I believe that the perceived special or unique role for us in the universe also has a link to the psychology that forces us to always assume someone/something is watching us—be it a god, or an alien, or whatever we choose. The idea that the wonderfulness of our being is only limited by limitations we create suffers because of all this.

But, couldn't it equally be limitless in awfulness? Yes. But, we have always known this. Let's use our logic. For example, as of writing this it is 2013. Now in seven years it will be 2020. We can assume a dystopia develops and work towards that, or we can assume that we can maintain current trajectories and favour this. But, remember there is no one watching, there are few relevant limits, even now; and we can make it as utopian as we can in that period—and come back and make it better again thereafter, and so on. Because, when you think of it, when you lay out these three options…why wouldn't we?

I believe that this positive possible side of the increasingly limitless potential must be held up. Many core decision makers could do with re-evaluating what they perceive to be limits and really, having appreciated technological development's acceleration, they need to up their game. 'Things could be worse' — that is no defence (neither is 'they were worse before'), but as long as they can be better, what else, logically, should we strive for?

A lot of this Book 3, though, probably veers somewhat away from the main concerns of the book as it becomes a bit of a polemic; just bear in mind that these concepts are my reflection on some of the implications of appreciating our physical, malleable natures, and the reason we have found it so hard to explain ourselves and the limit of any real limitations to what we might do. These reflections may not be agreeable to the reader, but the context (or Book 2) is the more important element from which they are to be viewed.

From Beyond the Subjectivity Trap

Within the subjectivity trap we are mysterious, soulful, and spiritual, maybe even a deity. From outside we are the same as everything else. We began questioning ourselves and even our ability to question as a result of the particular evolution of our communicative functionality, and we navel-gazed our way through history and philosophy, we created millennia of infantile belief systems and certainties, all thrown down in their turn. In reality (or closer to an agreeable position to take on our shared surrounds) we are not all that central to the universe. From a more positive slant we may point to the attractiveness of the notion that the concept of 'impossible' is just a useful tool to us, i.e. as our ancestor edges towards a cliff, he is 'certain' that he must not go over as he will die, and will state that it is impossible to fall from the cliff and live. It is a development of the most basic instinct of self-preservation. In fact, *impossible is nothing, or as close to it in light speed terms as to not make much difference to our loftiest goals*. In order to avoid dying as you fall off the cliff, just use nanobots to enhance your biological self with a titanium-like shell — or get a parachute.

We can, as is pointed out in Book 2, believe in just about anything. The problem is not that we believe in a particular thing, but that we believe in belief. Uncertainty and curiosity are the logical (least illogical) human standpoints and also breed openness and understanding. It is the lazy and unhelpful option to become certain, to become conservative. It is illogical, but it feels right. It is a social maturing process where one feels, 'I used to know so little, but now I know it all'. This sits counter to the truth. It is telling really that great minds, be it a Socrates and his wisdom in how little he knew or an Einstein and his childlike enthusiasm, seem to know that the *lack* of certainty is the key!

Implications

Let's tease out the implications of the physicality of humanity. This is to an extent a re-learning of how we can interpret almost anything. How do people watch silly TV shows — how can some people believe in war or violence, how could Nazi soldiers watch young children enter gas chambers? The liberating and terrifying truth is that all of these things clearly are possible. In fact, there is a massive spectrum of human behaviour. We could all agree to head out at an appointed time and kill the person on our left, ensuring that if someone hasn't killed you within a minute you would then kill yourself. This is unlikely, I am glad to speculate, but to truly liberate ourselves (or grow up, collectively) we must accept that possibility, as well as the possibility that we strive to live indefinitely in ever improving circumstances. There is room for fables with children who are developing mentally, but societies need to stop being childlike at some point, and morality essentially drawn from 'you can't do this' or 'you will be damned if you do that' and so on isn't helpful. It should be sufficient to show in simple terms that ideas of mutual benefit and cooperation have underpinned what is best in our evolution. The pivotal role of empathy and the reality that, though clearly we are individuals, we are members of societies, or but co-dependent creatures all part of the human race and also, indeed, part of this one biosphere.

Limitless: Relevant?

Are we insignificant, without a purpose, doomed to isolation and pointless pondering? Yes, and... No. If you extend my arguments regarding the lack of any destiny and purpose and you incorporate the vastness of our solar system or our universe compared to us and our scope, then it is hard to escape this outlook.

However, there is another way to look at our species, and that is one of unbridled (perhaps negative) potential. The argument I would present against our complete irrelevance in the context of what we know is this. If we spent all of our best efforts, for an indefinite period of time with the very best in cooperation and mutual purpose, and perhaps it would take thousands of millennia, all the while concentrating on one thing—to destroy our own solar system—I believe there is no way to reasonably argue that this is beyond us or our distant descendants. Destroy the universe, why not? Since we started messing around with flint and then telling each other how we did something, we did to some extent become different to everything else in the universe that we can see and, most likely, will ever see. What's that if not unbridled potential?

I think it is vital to bear in mind that no one else is going to say well done to us on this (the need for species-based reassurance is deeply ingrained) but what are also deeply ingrained are our limitations. Many of our cautionary memes, like the walking-under-the-ladder logic, are reminders of our basic evolved drive for self-preservation, and they're very useful; but others are deeply counter-productive, protecting us from our own potential.

Of course, there is potential to interpret the lack of fundamental behaviour-governing models, be it of a religious or any other nature, stemming from this overall, gradually occurring worldview revolution, as dangerous in the extreme. However, as said, it looks as if neuroscience is going to force our hands on this soon enough in any event. How long can we deny the functionality of what we see as our consciousness, our free will, and much else when experimental evidence continues to reinforce the functional biological make-up of our brains? It will no longer be enough to explain that you shouldn't do this or that because God or something else said so—we will have to explain, particularly to our youth, why there is not really a logical exclusive subjective perspective in social constructs and why in the end cooperation is our greatest tool. We will know and show instinctively without anyone's decree how love is the key drive for the vast majority of people (allowing that, unfortunately, it is perhaps frustrated and warped at times). Much of what we have claimed via superstition or religion or social constructs are clearly identifiable as warnings—much like the ladder example above (do not kill someone or you will go to hell)—they are a way of saying we as a society/species appreciate that cooperation and empathy are our key strengths and we wish people not to behave in such ways that will ultimately be the downfall of all people.

There are a few real impacts of teasing out the implications of all of this on our behaviours, it strikes me. Some of the items we could do without would include—just off the top of my head here—insanity, certainty, the existence of a 'right' way, truth, conservatism.

Not Right in the Head

If we really do comprehend what the brain is about we will understand what Metzinger talks of with his Phenomenal Self-Model, describing the brain as a virtual engine; and we will know that by definition there is no correct, objective interpretation of 'reality'. It is mere interpretation.

Does this mean that psychosis doesn't exist? Well, of course not—this would be a simple inside/outside subjectivity trap problem. We really can't go around speculating that it is all an illusion at all times—so why would we bother. This would be akin to being in the cinema for the greatest interactive movie of all time and continuously noting that this is only a movie and not real. We should, however, find a renewed fervour for our humility, and the behaviours of the crowd (or normalcy or consensus) should be seen surely as useful but not some form of definitive superiority.

When we look at this we can see that *disproportionate* levels of seeing or perceiving things as terrible or terrifying or 'not real' could be manifestations of depression, anxiety, and delusional disorders.

The little journey through Cognitive Behavioural Therapy can be reviewed again here. Why is it that we have such varying methods of dealing with these things? Why is it that sometimes talking about one's life (psychoanalysis) and other times mere habits of thought and behavioural associations (CBT) can work to alleviate some of these difficulties? How is it that sometimes medication assists also? How mysterious. But, not really, if we understand the machine as generating these 'internal representations' then the discursive or the cognitive habit-forming is similar in outcome, but different in method; medication dampens a set of activities with the intention of dampening the negative or troublesome activity within this set; with today's medication this is very much sledgehammer rather than scalpel!

In any event, promoting any forms of activity among many members of society, and not just the young, to breathe better, to have their awareness relaxed, and brought back from inflammatory responses of fight or flight to misdirected circumstances, should be widely supported across society — and understanding that everyone has mental health and, like physical health, it can be worked on and improved may be a good starting point. As stated, also notions of a dividing line between sane and insane should cease.

So we highlighted that 1) we are not in any way special, and 2) we have near to endless potential.

It seems that talk of potential often throws us into a collective insecurity. In fact, in many ways, it feels as if we are in a stage of infancy of our collective and individual thinking. It always seems as if we are imagining someone else to impress, be it a deity, or an alien, or some indescribable thing. It seems as if the greatest bad news that humanity could conclude is that there is nothing else, there are no preordained rules, there are absolutely no reasons to continue with physical suffering, death, and impoverishment, and so on, other than choosing to do so or not choosing to work on resolutions.

We have a tendency, too, of getting extraordinarily excited and worked into a frenzy over something new in what might be said to be an immature way—but we are at each new decade, in fact, very young. I mean this literally. How many people lived through the experience of the Wall Street crash in 1929 and are willing to base their decisions on that experience in the current economic turmoil? How useful would it be if we could collectively maintain our individual wisdom? This seemingly intractable overreliance on youth and inexperience will be looked at a little later.

What's the Meaning of Life?

So, from beyond the subjectivity trap what is the meaning of life? This can be a little scary, though hopefully we will see it is the change that scares and not the outcome. Currently, when we ask that question, how satisfying are the answers? How much do they help humanity, in truth? There is mass delusion here—and attempted ignoring of the tragedy of our own demise. Well, that demise is something I want to address in the next section—and it is going to require you muster up everything from the journey thus far as you will recoil at the positive possibilities that stare us in the face; but for the moment let us stick to the meaning of life. Now, one interpretation of the subjectivity trap is that there is no meaning of life. Personally, I think the better take is that the question itself is wrong—as again we look for something other than ourselves to assist; this childlike prevalence. The question should be, what meaning are we going to give it?

What about the morality, though? Maybe you can see the progression from early communication, even from early civilisation, and the benefit of being able to explain things, and the sense of there always being another authority, and gods of the unexplained thunder and lightning, and so on; and heliocentrism, and eventually just a God of man. And then you agree that perhaps it is more likely that we are a part of the vast universe we inhabit and not separated from it by our intelligence, and that as with everything else in us and everything else in the universe our consciousness merely seems inexplicable by mere physics, chemistry, and biology—not because it is, but because we ascribe attributes to it and an individual existence about it— which it itself dictates falsely.

And so, as it is our conscious 'I' that wonders of our conscious 'I', ultimately the thing being measured keeps coming up wrong because the measuring instrument is imperfect… So, say eventually you come around to all of that… but you have arrived there with the same conscious 'I' kicking and screaming for its own survival—and if you get that far in this mental journey (I am worried you have come further than most!)—I can clearly see another reasonable option then being promoted in your reasoning.

Essentially this would manifest itself as the following internal monologue: *'I get it, I just don't like it. There is an emptiness to it. Whether it is religion or something else, I want my life to have a meaning. I want to believe in something bigger than me. The fact that we are just certain forms of collections of atoms, and when the pattern loses its integrity we are no more and never to be again, is just too sad to bear.'* Again, I understand this in a particular paradigm, but in many ways the very best news we can infer in this text is left to the last. Please just let me park any possible desire to become an agnostic for a little while.

Because, firstly, there is an associated issue that can also arise here when people are presented with no neatly packaged meaning of life. This is the seeming lack of morality that might follow.

I will make one observation here. People who form these opinions now—atheists—particularly those who strongly advocate atheism (by this I mean not just people who don't believe in anything particular—but strongly believe that there is nothing), people in other words who see themselves as a brief pattern that will decompose along with everyone they know in time, are still (somewhat incredibly) scoring higher than their counterparts in many measures of morality. Surely the assumption would be that the despair of the reality of their predicament would lead to mass acts of violence against life, against one's own life, against everything?

How is it, then, that they do not do this, and that, as I said, their sense of morality is often deemed higher than that of those who profess religious beliefs? It seems a wonderful reflection of our predisposition to kindness, doing the right thing, and empathy. In fact, it might well be proposed that the position of 'being' religious itself has a deep-seated self-delusion which is a negative contributor to the psychological well-being of the individual. (Hopefully you see this comment as not contradictory to the main thrust of the Subjectivity Trap, but an observation valid from within.) There may well linger some doubt in the professor of a faith which rankles as they contemplate their mortality.

In any event, I am hoping to propose that that is almost an irrelevant point for our purposes, other than to highlight our incredibly strong starting morality. The possible within the context of this discussion is much, much better!

Longevity

No piece of matter emanates from a higher power, and all, obviously, can be manipulated.

A major outcome of escaping the subjectivity trap and realising our cosmic insignificance, isolation, responsibility for our own actions, and limitless potential is the ability to challenge the inevitability of our biological demise.

Let's remind ourselves again what we are. Let's look at our understanding of our biological self. Our entire entity is a tremendously complex thing, and there have been many, many lifetimes of dedication to understanding a wide array of physical ailments and infections and genetic defects and so on, with little progress in many areas. There are 100 trillion cells in our being, of which *just 10 trillion are our own* (another interesting observation on our assumed complete integrity of unchanging identity) — the other 90% forming the bacteria that line up along the many metres of knotted up digestive tract. There are a number of systems that include: the musculoskeletal system, cardiovascular system, digestive system, endocrine system, integumentary system, urinary system, lymphatic system, immune system, respiratory system, nervous system, and reproductive system. Bones and muscles form the musculoskeletal system (as I am sure a quick look at the name would tell you), providing solidity and mobility. Skin, hair, and nails form the outer protection from the elements of the integumentary system. Organs such as the heart and the brain form the centres of systems such as the cardiovascular and nervous system.

These systems are all quite clearly interdependent and so difficulties in the functionality of one will most likely have a spill-over into other areas. We may believe that there is some form of perfection going on here (but, again, we would be a little biased!). There is an incredibly complex system that has evolved from very simple fundamental parts by layers building on top of the other.

Perhaps it is worth reminding ourselves again here what we mean by the subjectivity trap. We are what is described above, made of the same stuff as is everything else, made of stardust. From a non-human perspective, the fact that we can say 'I' does not mean nearly as much as it seems to us saying it. If we can open our perspectives from this bias, we see that our saying 'I' is just one more thing that is occurring in a universe of things.

We have evolved and through that evolution we developed communication, and although we can communicate this does not justify separating ourselves from the universe we are so clearly within, nor using our mind's musings to suggest that the mind is separate from that which can only generate it, nor, in good faith, dreaming up very illogical conclusions to retain these myths in the face of mounting clarity of the inner workings of us.

There are an inordinate number of neurons and connections, but there are also vast numbers of sensory nerves and muscle controls and bodily instructions that revert back to the neural bundle that is our brain. I am sure large portions of its most recently developed surface will show themselves to be offered up to the act of outward communication and internal unstated communication, or thought. (While I must apologise for the repetition, it's important to stress that new concepts to any reader need reinforcement.)

Returning to our biological evolution, then, clearly it has and is serving us very well, but the idea of the impossibility of changing the system or certain parts of the system is illogical and, if you are talking to a cystic fibrosis sufferer for example, can be a very frustrating starting point for engagement.

The first battle cry against any mention of transhumanism (using scientific and technological methods to improve the human entity) is that it is 'unnatural' in many instances. I apologise for picking another 'sacred cow'. But if we are to avoid intellectual laziness, again, from everything we know this is just an irrational perspective. There isn't any underlying principle that says that there is any such thing as natural and unnatural.

Our perspectives often give us this sense of the world we find and the dangers of certain interactions. Of course, this is logical. If we kill all members of a species (not unheard of) there are knock-on effects, but this is confusing doing something with a negative outcome and a 'law'. The species may be made extinct without our intervention, but aside from that, the reality is that we are immersed in this biosphere. Literally, at every instance, unimaginably large numbers of living entities enter our entity, more again are full-time residents, and we shed practically our whole beings repeatedly over time. To hold this notion of our unique perfection that simply cannot be tampered with is wrong, but worse than that it is dangerous, as evidenced by many groups, mostly religious ethics groups, that are so sure in their defence of a 'natural' or God-ordained order of things that they take the unenviable (and I am sure difficult) stance of opposing parents of children with mitochondrial diseases and appalling quality of lives in arguing that research on genetic engineering (even of the next generation of children) is wrong — the idea that it would 'open a Pandora's box', etc.

Transhumanism is often a term that elicits fear. Sometimes it is seen as an animation club with a childlike and/or extremist's pathway to super hero status. It can also sometimes be represented by those with a desire to be anti-human and so change humanity fundamentally; it has even been taken up by some who have a particularly unpleasant political approach where the future is one of human advancement through technology by coercion. By strict definition, though, what is in effect being discussed is a mere evolution and refinement of the steps that support our entities that we have always taken. It is using technology to improve our condition.

Essentially, an eyeglass could be seen as transhumanist, as could a hearing aid. Cochlear implants and retinal implants are actually integrated into the nervous system. This would have been seen as impossible—and wrong—in many ways just a few decades ago, but try telling that to a parent of a young child receiving a cochlear implant. When we ingest food we are altering some of the structures of our entity. So we should not be quite so precious in theory, though in practice we should probably be more so, as we ingest and encounter so much that directly damages the system itself. Yet, many will scream 'don't do any of it', not, 'help, but do it carefully'.

It may seem ridiculous that a system that can drown or be burnt or frozen to death in relatively (universally or even just in earthly terms) benign circumstances should need such cognitive effort to be seen as imperfect, but I believe this notion of evolutionary perfection to be a close relation of the subjectivity trap. Evolution is an amazing thing, but it is just a process. To tease out imperfections even from the enclosure of our senses, let's look at the laryngeal nerve, which loops around the aorta and so travels a pretty nonsensical route in humans but, as is often pointed out, takes a ridiculous route up and down the neck of a giraffe. It is often cited as evidence of evolution (somehow this is still required in some circumstances) as the loop taken up and down the neck via the aortic arch is a re-route of a few inches in humans. But in the giraffe the nerve is an incredible 15 feet long. Once the nerve was formed, in evolutionary terms the development of the mammals from preceding fish-like creatures (and in the case of the giraffe the extending neck following the leaves of the higher branches) did not afford the opportunity to correct this re-route, and so the nerve travels such a long route to bridge such a small distance. It is particularly inefficient and used to argue against 'intelligent design' proponents. We have evolved under principles of natural selection and mutations, but we are not perfect, of course not, by even our own reasonable measures.

We are not perfect, but we are robust structure-/pattern-retaining entities. We are not, for example, a weather system which can manifestly change its substantive parts based on haphazard and distinctly unpredictable occurrences.

We are so robust that despite walking around in an atmosphere for ninety years, ingesting mind-blowing volumes of chemicals, digesting an incredible amount of protein sourcing foods, dispersing mountains of waste, and, indeed, growing, suffering injury, adding layer upon layer of experience, we still typically claim to be the 'same' person. We manifestly are not the same person. This is a part of the story we must create as part of our societally learned self-identification. I mean, from childhood to old age we are incredibly different; in fact, what remains and is carried unchanging (well, slowly changing) is the expression of our DNA code; and this in itself is testament contradictorily to our misplaced cohesive sense of unchanging identity, but also our robustness as entities.

We now know that practically all our cells die an awful lot faster than 'we' die; there are some that die in about five days, some that die every fifteen years, and they average somewhere about the seven-year mark, with a tiny percentage continuing throughout a lifetime. The cells that form our skin emerge up through layers until they are at our outermost surface and then fall away. Imagine if someone told you that huge chunks of your car (the vast, vast majority—with some parts being replaced every few days) will be replaced by very similar, if a little more degenerated, components over decades, created following the same instructions and made of the same materials! Would you consider it the same car? Would you pay the same price for it? What remains at all? If only there were a uniform code—something which underlay all else in the body... well, what do you know!

We are, I think, just leaving the infancy of our understanding of the genome, but we are at the beginning of a journey here that has some pretty astounding consequences. As of writing, already we create creatures that illuminate in the dark by activating certain gene formation interventions. The pace of developments here is gaining tremendous momentum—as complicated a system as it is, it is just a system, and with massive computation and the slow march towards sharing knowledge widely we are chiselling away at the unknown all the time.

I believe that one of the true ironies is that when one advocates that ageing is not in some way necessary (and if you have been reading thus far, you should not see anything as 'necessary' or as an unbreakable law) this is often met by those disclaiming this type of thinking as egocentric: why can't you just accept death as a natural process? Look at all the people who have died before you, there have always been charlatans who claimed they could defeat ageing but it never happened because it is impossible... etc. When we dig a little deeper, though, we see that one of the reasons that people are striving to hold on to this point of view is because they think that they are too special to be malleable to this sort of interference, or that they are too special to be in any way replicable.

Really – Fighting Ageing is at the Very Bottom Rung of the Possibility Ladder

There are many people working on extending biological longevity all over the world, and, in fact, there have been a host of different attempts to pull people together and create global organisations to maximise fundraising, etc., but essentially people need to see that ageing is not at all inevitable.

Regarding the 'there have always been those who claimed they have found the secret of youth' argument, it must be understood that we haven't had any idea what we were looking at really until very recently in human history. Of course we do not have the full picture now, or we would have resolved the degeneration problem, entirely, already. But it is important to note that we didn't even have a microscope until the sixteenth century. We didn't know the recipe of life or our DNA until exactly 60 years ago; we hadn't sequenced a genome until a dozen years ago, approximately.

Please let us accept, though, that we have—repeat, *we have*—not only creatures such as field mice that live much longer with treatments but, via biomarkers (things in the biology that show us the age of a creature), we have already rejuvenated; a proof of concept that would be seen as impossible, ridiculous, I would argue, right throughout human history until now; and rightly so, because it was.

Much of this needs to be understood at scale. If you were studying the brain by its size and shape because that is all you could see, you wouldn't know very much. In many ways much of medical history could be likened to this because we were treating something at one scale while its core mechanisms were occurring at a much smaller scale.

Sidetrack for Survival

As it turns out a whole bunch of 'scientists' proposed this study of brain size and shape... It was called phrenology. It was wrong, dangerous, and unfortunately taken incorrectly along with a number of other scientific 'breakthroughs' of a century ago. As an aside some people point to the dangers of scientific understanding with a broad view that says, with Darwin particularly, and a broader rise in the significance of science, we have a contributor to many violent beliefs and their ensuing tragedies, such as fascism, in the twentieth century. The problem was not the discoveries, but the intellectually lazy reporting of them and the ensuing understanding or misunderstanding. Personally I would suggest that 'survival of the fittest' turned into the most dangerous concept emanating from science in human history. But, of course, it was completely misunderstood. It was seen as an excuse for one group to claim superiority and believe that the exter-

mination of another and propagation of the superior would be beneficial. Aside from the complete lack of understanding of real genetic differences and the scope of time for lineage change (or fundamentally through just 150,000 human years), it is practically ridiculous to talk of anything other than the human race being just that – one race. Aside from all that (and who decides what is superior in any event) it just wasn't understood what was meant by Darwin to be the 'fittest', that this referred to fittest to the environment.

There is a nice oft-cited example of the peppered moth in Britain, which was found in two varieties, light coloured and black, prior to the industrial revolution. The degree of camouflage offered by the trees they lived on was very significant to the number of birds of prey they encountered. So at one point the light coloured moth represented 99.9% of the population. During the industrial revolution, owing to a lot of smoke and so on, the trees' barks darkened; the population of the black moth then peaked at 98%. Of course, after WWII the British government enacted environmental protective legislation which greatly reduced the damage caused to the bark and which returned it to its original colour – and you can guess what happened with the moth. <u>This</u> is what is meant by 'fittest'. It is natural selection, and it is simple mathematics; in other words fewer light coloured moths were having progeny because more of them were dying, and more of the black moths were procreating as they were surviving, due to this fundamental change in their environment.

Whether in the horrors of war or the unpleasantness of corporate colour it seems that the misunderstanding continues in many quarters, even if it is reframed as benefits of competition, but it seems to ignore that it was, most likely, the very empathic drive and the ability to communicate, collaborate, and mutually learn and record what was learned for future generations that provided our strongest tools in the broadest environment and perpetuated our success as a species.

The Longevity Bus

Where is the smart money on the breakthroughs for longevity? Well, first off there really isn't enough research underway, nor are there enough people working on it—these are really regrettable hangovers from the history of wolf-crying on ageing, the ingrained necessary acceptance of this inevitable tragedy, and the resistance to the possibility of significantly affecting ageing, however clearly evidenced and theorised.

But, we must accept that the hard truth is that, despite the very strong evidence of logical analysis of our selves and our surroundings assisted by highlighting our own lack of a unique position in the universe, every day approximately 100,000 people die of ageing-related diseases.

By the way, no one dies of just ageing, it's just that your chances of developing these diseases rise each year—but, you do die of a disease.

Returning to the smart money, there are three competing theories as to why we age. Sorry, that is not necessarily accurate, there are three theories that I am going to discuss as to how we age.

So, there is one school of thought that effectively states that ageing is pre-programmed, as we develop from zygote to fetus to infant to child to adult we are also pre-programmed to develop into decrepitude, and so this is inbuilt in our original design.

Another is that we grow to become adult and function successfully as such. Then, however, small amounts of damage occur to our systems and at the lower levels these come from cellular damage, and it is a build-up of damage that stops the cells looking and working like they have done in the past and, as with any system, damage on top of damage eventually gives rise to very bad situations. As an example—we have mitochondria which act like power houses to the cells' activities and take energy we ingest and bring it to work for the cell itself. They produce waste in so doing, and over time this waste starts to accumulate. The SENS foundation's Audbrey de Grey has challenged people to expand on the seven broad categories of damage that he outlines (even offering a prize to anyone to reasonably show another category of damage—a prize that remains uncollected over many years). The seven causes all have potential or theoretical therapies underway. This is all under the broad idea of the 'Damage Theory of Ageing'.

Another concept around the primary cause of ageing that I would like to highlight is that there is a tip end to your DNA strands called the telomere that is chiselled away at each time a cell divides and the two strands attach elsewhere for replication and to create a new cell. The accuracy of the replication process is damaged gradually as the telomere is shortened. And thus much of the damage, including particularly DNA mutations, emanate from this loss of fidelity in the copying cell. There have been studies showing the connection between telomere length and longevity in many animals.

Whether it is pre-programmed and environmental epigenetic changes or accumulated damage or telomere shortening that show themselves as more important than the others in terms of stalling the degenerative ageing process, we need to be honest with ourselves that it is really only a matter of timing as to when we have viable treatments to resist the loss of functionality currently inevitable with accumulated ageing. It may be hundreds of years off, as a few researchers still state, but almost no one in a scientific field maintains that it is fundamentally impossible to halt the ageing process.

There has already been success with anti-ageing therapies for other creatures, including some very complex mammals such as mice. The effects have ranged from a 20–30% increase to a ten-fold increase in nematode worms' lifespans with certain treatments. This is not impossible. It is not simple, but it is most assuredly not impossible.

If there is a part of you that says, 'well, no it might be impossible', 'how do we know', etc.—please force yourself to make a logical argument that questions why over the next 100,000 years of studying, mapping, and just getting current computer algorithms to run treatment simulations, there is in this universe a valid reason why reduction of the negative aspects associated with the ageing process might not be achieved to substantially extend healthy lifespan or healthspan.

We might claim that in 100,000 years we will be long gone—but there is the minor point here that, if we work on it now, we are still perhaps working to save what is in today's terms

100,000 people that die every day from the ageing process at that future point. Then there is the much more realistic look at this. All best indicators are that this is a process that will take several decades. There are some indicators from current progress that early therapies are even much closer than that. We must factor in here, as well, that the 'human complexity bias' has a major role to play. In truth, we can have the timing debate forever without funding the excellent underfunded work that is taking place now. What leads to progress is many people working together on these topics, erring and learning and progressing. This all takes time, time which would be shortened by having more and more effort being applied by more and more people right now, or — in other words — providing money for research.

Recall what we said that no one dies of ageing — as your body degenerates over time you become more and more likely to succumb to any one of a plethora of diseases. It is important for us to see that a vast array of the causes of our deaths, neurodegenerative, cardiovascular, osteoporosis diseases, and others, all clearly have as their underlying cause inextricable links to what we view as the ageing process. When we study these diseases (as we do now) we, indirectly, study ageing. Also there is a very close link between the degenerative ageing process and the growth of the non-dying cancerous cell. And this absolutely tragic plague of today's life uses an interesting means to attain its 'immortality' — it produces the telomere-strengthening protein telomerase. There is a very close link between the mechanisms of a cancerous cell and the ageing process, which again means that indirectly those that study cancer are gathering data on ageing also.

So why don't these people just say that they are studying ageing? Because we wouldn't give them any money — because we all believe (or at least the vast majority believe) that trying to halt the ageing process is silly!

Shouts abound to the effect that there have always been charlatans and snake-oil merchants purporting to have found 'the fountain of youth' throughout history — which is a mere reflection of how much we dislike the ageing process. There is often a venom to these arguments, which perhaps hints at an underlying wish not to have the ultimate of hopes dashed. Perhaps if the scientific realities were presented with the full spectrum of opinion included, at the least people may venture to hope within a reasoned context. This hope may well have many psychological benefits.

Just to turn to the word 'immortal' and also to 'immortal' cells. Organisms are made up of numbers of cells. There are many single cellular and a few multicellular entities that do not die from what is known as senescence. The sperm line cells in humans do not senesce. There is a theory that the somatic or non-sperm cells do, simply because the sperm cells are more important to continuing life through progeny, and this high maintenance of functionality requires a few more resources, and so it is simply an allocation of resources. Planarian flatworms' telomeres never shorten and they also never senesce—they do not get old. Freshwater creatures called hydra and a genus of jellyfish also show no signs of ageing. Bacteria colonies are also 'immortal'. I keep using these quotation marks because it is important to note that all of these creatures, and humans if organisations such as SENS in the UK and the Institute of Aging in US and a host of agencies in Moscow and elsewhere get the support they deserve, even with a perpetually rejuvenating body could still be poisoned or killed in an accident and so on, as frequently occurs with the immortal hydra. And so immortality (as defined by being unable to die) is an aspiration, never an attainment.

Religion

God

There is immense complexity in a human, and also in the extraordinary layers of societal constructs that we have—the self-evident need for rules and regulations around criminality and guidance around certain behaviours in myriad forms of retelling and conveying lessons learned through the individual and

Chapter Nine: Beyond the Subjectivity Trap 145

collective experience are all primary functions of societies, and often of religions also.

The best people I have ever known were and are persons of advancing years, some of whom are no longer with us. These, in my experience, took much of their wisdom, their social learning over long lifetimes, from religion. I do feel that these are persons who would have been very much the best of people in any event, if they had never been indoctrinated so, but we must acknowledge the benefits of moral grounding; in some of the modesty and self-sacrifice, in the care for others that can be seen as emanating across the religions. Undoubtedly the religious figure plays a very positive role in many cases when a tragedy strikes a community and acts as a focal point and source of strength.

There is a need, undoubtedly, to retain so much of the accumulated and handed-down morality and societal support. In a hypothetical future post-religious society it would be imperative that the supportive role is not extinguished.

Of course, on the negative side of the religious *modus operandi* is the idea that some unearthly entity gives any individual or group of persons a notional authority or bequeaths on them a greater likelihood to be more right than others! The dangers to the psychology of having a superiority complex reinforced ritually are painfully clear through religious orders in recent decades, in my own home country and elsewhere. Religion is therefore a mix of good and bad—it is human, I guess.

We must ask, supposing there is no God, where does the construct of God and the character of God come from? This non-religious perspective of the history and evolution of gods and religion is an interesting thought experiment also. It seems obvious that, throughout human communicative history, religion and supernatural beliefs were mechanisms by which individuals and groups could maintain certitude in the face of things that were not understood. We can see gods of thunder and Sun and a whole range of elements that at the time were mysterious. The gods we are most familiar with in this era are those of man only, and perhaps it can be inferred that man is seen by many as the last unexplained phenomenon, and so last to need a god.

In a worldview that accedes that because of subjectivity we can know, in an objective sense, nothing with certainty, but rather only with usefulness and best consensus among humans, then ruling out an existing God is also wrong. It is also oblivious to the comfort and identity many take from religious beliefs. So, personally I would never suggest that someone's God is not real *for them*; I would merely question their certainty in that God's existence.

Interestingly, we might ask, if God does not exist, where does his/her character come from? One assumption would be that the behaviours we have attributed to our gods are an expression of how we feel we should behave ourselves—God is inside you! Another assumption is that the omniscience and omnipotent but benign and, indeed, empathetic and loving nature of how we define God is also where ultimately we wish our destiny to lie. In God we express what we wish we were, or what we may become. A little like science fiction it may be seen as a pointer to our behavioural aspirations. From the notion of and expression of God there is much to learn. Therefore if we ever see ourselves evolve to a point where even time is at our command then it would be that God would exist now, even if God was, in fact, a manifestation of future man. Perhaps one could then add another to the mix of free will paradoxes that 'God' would not interfere in 'Man's' journey as Man eventually becomes God. And so, Man is made from God, because God is made from Man. In any event, this is doffing a cap to reassuring conjecture!

When we consider this history of man through the prism of there being no God, then one views religions and religious figures from this perspective and we can see errors in theory—we see some really negative and hateful ideologies, and we may look at many such ideologues and see them within their time as expressions of people trying to stop something they didn't like or understand by claiming that God does not approve.

Of course, this is one side, and indeed perhaps the smaller side. What we see in the majority are messages of understanding and cooperation, and forgiveness and love… Ultimately, what we see then are, in total, very human attributes indeed.

What is the impact of putting away childish notions of our special position; well, two I can think of easily. We must accept the fallacy of certainty of anything (and yes this applies to those who are certain there is not a God), but we must mostly, hopefully, take the implications for humility. It is the position that someone can believe in whatever they wish, and whatever perhaps keeps themselves and their families happy if they so wish, but they should, in keeping with appreciating that whatever they think is subject to the vagaries of their own fallibility, offer opinions only and not force what they believe to be 'the answer' on anyone else. Yes this, of course, applies not just to religion. One observation further, though; when longevity in its various proposed forms begins to bend the stranglehold of inevitable physical death, I wonder what will happen to people's religious beliefs if they are ever offered a viable alternative!

Politics, Society, Morality, and Stuff

Let's cut right to the chase here and look at just how impossible it is to explain some human behaviour. In 1942 in the city of Lens 4000 French kids were put on holocaust trains. The trains pulled up to Treblinka. Treblinka doesn't even bother with housing and working as with Auschwitz and other predecessors. It has the facade of a train station, including flower pots. The children are told they must have a shower and so they need to undress. The older children help the younger children to undress. Some of the younger ones are running around and playing. They will enter the 'station' and find themselves walled into a gas chamber and they will be gassed to death. They will feel ill and terrified and will scrape at their surroundings with bleeding hands, screaming and crying out for their parents.

It is valuable for many reasons to think through this horror. It is I feel dangerous to suggest that these things are so bad we must ignore them. It is valuable because their horror should not just be forgotten, ever. It is valuable because it reaffirms that almost all people at all times are repulsed to their very cores by the mere recounting. It is valuable also, though, because there were many soldiers who carried out these unimaginable horrors and who watched these events unfold without feeling that they were doing anything wrong. In fact, who really ever feels they are doing anything wrong? Who wakes up and says, 'I will do wrong today'? What psychopath on their way to their own massacre doesn't feel that what they are doing is right, justified, or important? It is valuable because we need to understand how this can be, and to see how it fits from the view beyond the subjectivity trap.

But there is an air of childishness about how we approach this generally, also. We wonder just how bad things can be — how badly it is possible to behave. Should we even think of these things, will thinking about them make them happen? Perhaps in Treblinka and elsewhere we have already teased this out. Things can be very, very bad. There isn't a limit set externally for us. The maturity to appreciate that we do or do not do these things and that they are individually and collectively our responsibility would certainly be progress. Logic again dictates that we do not wish or need and should not countenance allowing for circumstances to develop where they can occur again. The psychology of dehumanising any group must be resisted as much as we would refuse to partake in the 'showers' discussed above.

We, I think, in the limitless sense, need to open up our shoulders here and say, look, we got it, the twentieth century gave some real clear indicators of how bad things can be, lesson learned, there is absolutely no reason to maintain those behaviours. Having an opinion on anything, ever, that suggests something like, 'well, maybe for this a child should die', is just wrong. It flies against what the vast majority of humans socially and instinctively feel; there are moral but also evolutionary

Chapter Nine: Beyond the Subjectivity Trap 149

reasons why our hearts tear when we even think about these children.

The fact that I described an incident with children adds emotion perhaps to these thoughts but, in truth, at what point in your teenager's development do you begin to feel, 'well, I don't mind if someone kills him or her now'? And yet, children are dying, as I write, from nonsensical adult arguments, as has always been the case. In the end, religion has been an exercise through which we teased out morality — and in truth our behaviour towards each other on a large scale has improved through the centuries. Moral lessons are essential to our educations, but we must highlight the logic of these things, the fact that one cannot in any meaningful sense have more claim to being right or be more worthy than another, and that we are social creatures who benefit greatly by treating each other well.

Another seemingly common reaction is to look for personal culpability and naïvely assume that what can happen in one group of people cannot happen in another. This is not only immature, illogical, and arrogant but, ultimately, it is very dangerous thinking. The idea that particular races of people are capable of something that others are not is just the worst form of commonly held opinion. Repeatedly we hear people make racist claims, discuss national or group failings. It is vital to bear in mind that humans are like any other group explicable by a distribution curve, and if the majority of a group of people are largely behaving in a certain way at a certain time and place then so would the majority of all of us. This is a fact, and a very important fact. Any scant view of human history screams just one fact to you — there is one race — the human race. And people have a degree of individuality (your own sibling can differ quite significantly from you) and a degree of universality (you can find a friend or soulmate anywhere in the world) that modern communications technology should just leave no one in any doubt about.

An associated assumption is that there is only one Hitler — of course, persons with incredibly damaging and dangerous worldviews borne of major discrepancies in their world interpretations exist aplenty. In most circumstances and societies they are cared for, not empowered. And the lessons of how a disempowered society can opt for this type of individual to lead them (or indeed why there is need for *a* leader figure at all) are of immense interest. One really shouldn't see these things as that unusual either, there are current manifestations evident to varying degrees all over the globe.

In any event, this all speaks to the subjective problems we can see filtering from assumptions. The amount of times I have heard whole tens of millions of individual persons referred to as 'they are all x' or 'they are all y' I can hardly count. It is appalling thinking, but it is also self-reinforcing. I see, daily, people in public positions talk of a particular action being typical of a particular people—there is an 'anti-logic' in action here. If another nationality were to do the exact same thing it would go uncommented. In fact, often it is interpreted as representing a completely different trait, because this is the trait already ascribed to this particular people. Anyway, I rant a little but I think we all appreciate this is just, again, lazy thinking and it can take us down very dangerous roads, and undoubtedly nationalism is the grandest of irrational constructs we have and are perpetuating. Nationality—a concept that didn't exist mere centuries ago—in light of the damage it causes, and the complete lack of support for its claims (particularly in a world that an astronaut's camera can turn back and gaze on), is as Einstein famously claimed 'the measles of humankind—an infantile disease'.

What of a world without the moral certainty of national superiority or religious ideology, though? Would we without these constraints not enter into a world where anything is possible and therefore anything and everything could and would go wrong? Well, these are not points to be taken lightly, but we certainly are not emerging from a pristine past, and although the idea that it is completely up to us can mean of course that we go on to make the world a dystopia because it is up to us and nothing and no one can stop it... But also, as it is entirely up to us, it does not have a limit in terms of how great it could be either.

I recently (in 2013) heard a reasonable sounding military and diplomatic analyst talk on radio about US foreign policy and — as if he were commenting on a football match or the weather — he claimed that China wishes to have 20-30 years without a war to finish its process of development. The accuracy of the analysis is not what I wish to discuss; rather, how it is possible for someone to discuss a potential war between two massive nations in such a blasé manner is terrifying and shows a world where these things are in some ways normalised. I heard other things that quite simply terrify me too; the rise of fascism in Europe due to the politics of small-minded self-interest and politicians devoid of imagination or any comprehension of the exponential growth of technological development, which has prevailed and left unemployment and financial disempowerment in some countries at rates of half of the youthful population.

I have heard comments based on the inglorious idiocy of racism become ever more prevalent in recent years. All the while the danger of someone going rogue, or a training accident between two major powers, or, indeed, the inevitability of greater technology-enabled weapons easily produced, printed, or 'grown' by the outliers, all strike a great discomfort. This sits beside the wonderfulness of the global population increasing their knowledge of each other — the appreciation that there may be significant differences between any two people (as said they might even be in the same family) and so whatever is held to be in common to particular regional groups or races of people are at most ridiculously insignificant and superficial cultural conceits.

Anyway, dismounting from this rant, again, what can we possibly do about all of these negative trajectories? Have we ever imagined a scenario where distrusting and local groups came together to live in relative peace? Well, yes, this is the creation of the modern nation state, with its arms of judiciary, political power, police force, and journalists; all pitted against each other in a self-correcting balance. It (mostly) works. It is vastly more intelligent than the current 'accident waiting to happen' international circumstance we sit in. If only we had international bodies... Well, they are there already, bodies such as the UN, the international courts, the UN peacekeeping force, and so on. Required changes, as extraordinarily complex and difficult as they would be, would constitute really having international courts that could call on the police force (or UN peacekeeping and global criminal bodies) to bring to task any global leader or body that was seen to act outside of the international laws—any!

Undoubtedly this would require a maturing of current international trust. It would also most likely require bodies to push for this by organising online pressure on politicians that are elected locally and nationally to agree to fight for this internationalisation.

This is difficult of course, but it is also a contribution to halting current global tragedy as well as protecting against future disasters. Situations such as Syria at the moment; if there are breaches (declared by representative judiciary from over 200 countries and respecting of the UN declaration of human rights and Geneva Convention and so on) of internationally agreed minimum behaviours then the leader may be removed at threat of a United Nations peacekeeping force (the international police force), also representing over 200 countries, and imprisoned if violations are proven. This would lead, of course, to the possibility of leaders of western states as well as others facing this fate. Greater international cooperation on so many areas of human endeavour in fields such as science and medicine, international crime prevention, and management of disease outbreaks and weapons production (also, importantly, macro-management of global finance and global business), and so on, would all be accommodated.

When the urge to dismiss such ideas as unattainable strikes, it is imperative that we try to resist the urge to give in and accept the numbing nature of tragic events. It helps to personalise. A documentary recently showed one boy of about 11 in

Syria tending ever so gently to the seeping eye of an old man hurt in the conflict, in an unquestioning display of human nature and the desire, which is clear in practically all children, to help someone who is hurting. A little later in the documentary we saw this little boy with his head thrust back and his mouth opened to an unnatural degree, aghast at the moment his death impacted. Before we dismiss others and ourselves of any naïvety, we should remember this; remember the lack of information our predecessors had, and try just about everything to end this awful stupidity! Keep in mind that there are no set limits to how good we can be, no one is watching. And appreciate history, of course, and the difficulties and risks of change; but, really, if we fully imagine back in time to the reality of our ancestors' short and painful existences, and also their very limited understanding of the magnitudes of their surrounds, we must cease to use their ignorance as an excuse to perpetuate our unnecessary arguments.

The Syrian event and the completely childish and unpredictable international reaction(s) to it should highlight how easily a tipping point of escalation can be reached, and how quickly we might find ourselves in major trouble.

Yes, some may argue, but isn't this naïve? Isn't this just the nature of international relations and big business and the baddies, etc.? And hasn't it always been like this? Well, no. Go back just a few hundred years and talk to people of nations we all recognise now and they would have no idea what you are on about, these 'nations' are pure fictions. The mantra of historical tribal, and then much later national, relations could be likened to the incredibly unlikely event of us now learning of another existing travelling civilisation on a distant planet. We know nothing about them. So imagine SETI (our space-trained radios involved in the Search for Extra-Terrestrial Intelligence) finding a communication intercepted in some intelligent manner out there in space. Imagine the debate on Earth. We would assume that if we knew of them they must know of us. We would be wondering if they might attack us. Of course, we also assume that they are wondering the same thing. In this case the logical thing to do is to attack. Precisely because you must assume that they will also come to that conclusion and so survival in this information vacuum with strangers behoves a first strike. These attacks will elicit cycles of revenge and accompanying hatred. This scenario is useful to describe how groups of people in the world of our ancestors for whom the Earth was the vast and still-being-discovered universe, and where there were found groups that nothing was known of, would have found themselves in initial battle. And the dark shadow that this history can cast.

Conclusion — It Is Here Now

Returning to our longevity discussion — so, what now then? Even if we accept the theoretically likely resolution to much of the biological frailty associated with ageing, is this all some academic exercise, or at best will these rejuvenation technologies come at some point in the future, somewhere between the most optimistic and most pessimistic projections, and so possibly when you and I are long gone? Or perhaps the rejuvenation technologies will exist in your lifetime but you will be involved in a car accident or there will be some existential catastrophe like the ones mentioned above that renders the phenomenon of biological life a thing of the past. Is that it then?

If we do accept consciousness as an illusion, but an amazingly useful one, for us — should we not try to keep it? I mean, this body we are in, this consciousness of ours which as we have repeatedly said is a part of us, it is a part of and sits in a very vulnerable thing. Imagine Einstein at the beginning of 1905 choking on some food — we know this is possible, don't we? Or the fate of the children of Lens or the child doctor in Aleppo!

Live Forever, Now

You are in your brain and body! I am familiar with groups who spend vast amounts of time working on the tasks that will allow them to informationally map the entirety of the human brain. This is a major computational task. There are, as we have said, somewhere in the region of 100 billion neurons with 100 trillion synaptic connections or points where neurons communicate with each other. What is processed into the subjective I we want to keep as well as other underlying elements? Surely not all of it is necessary to copy at cellular level if we want our 'I' to survive…

So, what if you have an operation and your appendix is removed, you're still you when you reawaken, right? What about your tonsils? What about a limb? We might say that you are different to before, but you are still you, not a different person, just the same person changed. Are there parts of persons' brains damaged by stroke and other events? Absolutely — connections can be completely severed and the person may lose control of one side of their body, yet, I think we would all agree it was still them.

I would guess some people might say, 'well, it is the memories that matter, if I awoke from an anaesthetic and I had no memory, then I would not be me' (while others would conclude that it would still be them — just them without their memory).

Here I believe we are hitting on the kernel of the issue. If someone close to you was in an accident and suffered brain injury and had complete retrograde amnesia — or they could not remember anything before the accident but they could form new memories — you would most likely take that outcome for your loved one ahead of death; and you would try by all manner and means to show this person their own personal history. If you said absolutely nothing to this person, they would go to the bathroom, look in a mirror, wonder who 'they' were, but still feel they were them (their internal representation would still be in action)! So what is it then that is this 'person'?

Certainly, it is not actually a continuous being. You change all of the time. You are different now to when you were a child, and you are different now, in terms of raw anatomy, to what you were two minutes ago. But what of someone who, through a major psychotic episode, suffers a deficit in their ability to interpret the world or loses major brain functionality in an event or accident — would we be comfortable in saying that this was 'not' the same person? One can make the point as above that the person, regardless of how different they are, if they stood in front of mirror, if, in fact, they were a completely different person, they would still say, 'well, I am me!' And by extending this line of thought to extremity we may see that perhaps there is some sort of logic in believing in reincarnation, and that, in fact, when we think about it truly, we can only ever experience the moment we are in, and so any worry for our futures, minute to minute or after death, is illogical.

Think about this. You can never be the you of the next moment in time. You can only ever be you in this moment. There is nihilism possible here, yes? We can then say, 'well, if I die right now, what difference does it make?' And from this perspective the answer, of course, has to be — none. But, and it's a big but, we are not really like that; we are essentially a creature who is designed, for the most part, to survive to the next moment. It is something we want (we might well suggest that this, coupled with the desire to find nourishment and procreate, is the underlying reason why we exist).

Death is, also, incredibly sorrowful for those who have formed bonds with us, have found the ultimate happiness in the vastness of their universe with the intimate love of another. This is an example of where being human and being logical are again shown to not necessarily be absolutely compatible. One of the reasons we don't walk into fires and do many other things individually and collectively throughout our evolutionary lives is precisely because of this central drive in our wiring — to prepare in this moment for the next, and so to ensure survival from moment to moment. So, illogical as it may be, it is central to being human to want to continue, to live, to be.

There is a desire to build a continuous story — and, sure it may be delusional to say I am me and always have been, because by any measure of what is me I am changed utterly, but also, as the subjective 'I' is in itself a grand illusion, we must see that there is some essence of us, that within the illusion cohesiveness of self is very important.

Chapter Nine: Beyond the Subjectivity Trap

So what are we looking for, then, to continue beyond our biological limitations (even the limitations remaining post-eradication of degenerative ageing)? We want all the stuff that is you, we want it represented, coded in some way. From the philosophical debates above we can probably assume we want something that doesn't require atomic-level precision of brain recording, and something that doesn't just reassure us by telling us that we can come back as a dog after we die and that a dog somewhere will sense being themselves!

Some who contemplate concepts such as Whole Brain Emulation, where the entire information that creates you is recorded until a point where computation can replicate it, ask 'if I am stored as a recording, and there is a cure to a whole bunch of genetic stuff before I am brought back to a biological form, or in another substrate there are advances in my processing speed and complexity, am I really still me?'

I think the unsettling answer is that to *that* you it doesn't essentially matter—to others it would; however, we cannot do this for the *next* you, we must do it for *this* you.

Essentially what are you then, are you real at all? I think the necessary answer to this, for us, is yes. And we are essentially important to ourselves, but we are not entirely comprised of what we think we are and our essence needs teasing out. If we accept that if you lost some of your memories—or say they changed somewhat, as they do over time all the time (we know that a memory is a memory of the last time you remembered it)—you would still be yourself... then what is this 'our self', informationally speaking?

Well, we know we are not (in a universe with diversity and some structure) that cloud over there or that rock. We are our limbs and our torso, we are our fingerprints, the particularity of how we see and hear, we are the broad infrastructure (though plastic enough to change somewhat) of our neuronal networks, we are our unique retinas, we are what underlies these things, the physical (and it's all physical) changes we have undergone.

It would be really cool if there were a set of instructions which pre-programmed our design throughout our lives (and while open to environmental change were robust enough to give the continuity that currently suffices), if there was something that held the recipe for us!

Eureka

OK, so I guess every book needs a special bit of a 'Eureka' or 'holy shit' moment; well, here comes this book's one. I am sure you have perceived that we are already blessed with the informational, beautifully compressed DNA we all carry. I would like you to take a couple of thoughts with you at this point, which are extraordinarily difficult to disagree with.

Firstly, when we lose consciousness (or have decreased brain activity) time lapse does not have as clear an impact. So, when awakening from a coma we may hear of a case where the individual's mind and memory are still lodged firmly in the twenty-year-old time frame that they entered the coma in the first place in. From sleep to anaesthetic to coma we can see that the greater the depth of unconsciousness, or lack of brain activity, the greater the loss of a sense of time lapse — in other words, we need to be conscious to experience time. We can fairly safely assume that when you are dead you are unconscious, or when there is zero brain activity there is zero time lapse.

The other consideration I would like you to concentrate on at this point is the ultimate understanding of the brain attainable over an undefined period of time. In other words, whatever the brain is it is contained in your head, it is a part of your anatomy; it is a closed system with a finite amount of information. At some point in ten years, ten thousand years, or ten million years study of this system will unveil its informational structure and integrity. The only criteria required are that people exist and that they pursue this knowledge for as long as it takes. Remember, they will be doing this work for themselves also!

Combine these two points. You lie on your deathbed. You offer as many representations as is possible of your informational treasure. This may be via brain scans, DNA samples, cryopreservation, video-diaries, and so on; even at today's levels and appreciating the retrograde amnesia argument (better your loved one emerged from an accident alive but with no memories but the ability to form new ones, surely, than death) it's possible to see that *you would die in the knowledge that you will awaken instantaneously*.

This is how it would occur on your clock, to your awareness, and that is all that matters on your deathbed. Your perspective will involve instantaneous reawakening. The degree of resolution of brain mapping may be very problematic now for total atomic-level precision but there are advances in the immediate future that may alter this substantially... *As of writing there is an article published about 'neural dust', which may operate at such a minute level that it can take up residence in the infrastructure of the brain without causing any harm and relay signals to a recording device; and the journey into high resolution and the breadth of neural activity recording advance seems to be underway.*

But as we have pointed out already, that level of fidelity is highly unlikely to be necessary for us to continue this illusory but useful sense of conscious existence. In fact, we might well point out that retrieving the DNA of the already deceased would lead, at the very least, to the retrograde amnesia standard of person representation discussed above.

This argument can be dragged out of quasi-philosophical musings and logical, but ridiculous, reductionism to the world we know and are familiar with. There is an organistaion we are familiar with here in Ireland, though I am sure it is international, called 'The Make-a-Wish Foundation'. I, and I'm pretty sure I'm not alone here, find it difficult to even consider what this organisation is about without feeling a swell of grief in my chest. The idea of a child dying, the idea of the children's parents, siblings, family, friends saying goodbye to this child as their biological essence fails, is just too hard to bear. It is the most horrible thing there is, surely. I beg you to consider, is it not worth trying at least to retain the totality of the information of what is the physics and chemistry of this child for future understanding and possible reanimation? I hope at this point of the book you appreciate where I stand on this—I feel it is just a reflection of our own ego-centrism, ironically, when we consider what we are trying to achieve here, that tells us we are inimitable. I believe that with DNA we have the thing that is you, or at least the instructions of you.

There is, of course, the necessity to accept the role of experience in the brain; and undoubtedly it is a highly plastic organ that can be affected by events. There is also a very robust element to its innate design though. We know, for example, if two newborns are exposed to the same experience, they are already predisposed differently and will interpret differently (their eyes will see differently, register colours differently, etc. and their bodies will feel slightly differently). But predisposed by what though? By their interior and total structure, which is encoded in their DNA?

While certainly our memories are incredibly important to us, they are loops of prediction that will be seen in the branching of neurons… The pathway of this conversation is complex, but it is unlikely that we operate like a computer filing system. Rather, there is a broad representation of an attractive person, for example, and someone who is taller than the other within this category (different eyes, seen on a TV, etc.) and is recognised as such. And by more differentials we recall the individual through the changes from categories.

Also, as the memories don't reside in a particular location but form a part of this narrative functionality with the potential to 'branch out' by associations and differentials, we must countenance that—however unpalatable—if I am represented via DNA, my retina is the same in this representation. And in terms of memories—is there really a difference in being informed (having it represented in neural patterns) that I saw the assassination of JFK in a particular setting; as opposed to actually being in this setting in 1963? These are difficult areas no doubt.

There are, as you can see, pedantic elements to this argument, but we need to zoom out and think about the robustness of the entity, how many things may occur before you would assume (in a generally accepted way) that you are no longer you! In truth it is a pretty exhaustive list; add to this the illusory and malleable nature of self, and the possibility of extracting in the very near future (with appropriate funding and endeavour) synaptic resolution brain mapping, and you must conclude that at this moment in time there is at the very least a really strong argument that we can look at the child on the bed with a very strong hope of seeing them instantaneously (on their clock) after they expire. This is the ability to say (with at least some merit — I would argue with logical certainty — but am aware it is very difficult to arrive at this point) to your child, *'I will hold your hand, I will be here with you, and when you die I will be there to hold your hand again, it will happen the instant you die, you will reawaken and I will still be holding your hand'*.

So, is this all possible now? No.

If I die right now, if you die right now, we will disintegrate and decompose. And that incredible pattern that was you, that took a whole evolution-time to create, will cease because no one is taking and keeping the information. This should be eradicated immediately, collectively we should create institutions that offer as a right to all the storage of DNA and the uploading of video-diaries and so on, plus the creation of what are being called mindfiles, where persons leave vast quantities of diaries, opinion pieces, thoughts, responses to questions, and so on — that help in the refinement of the fidelity process, perhaps. In Ireland now, the cheapest option for disposing of your body under the ground to rot in private runs to about €8,000.

Video diaries might be kept by the individual themselves, where they discuss memories or run through their thoughts and feelings/responses to questions, for indefinite storage until neuronal representation is computed. As well as this, or in place of it for the already deceased, the reanimated may absorb or watch reports from those close to them on events after their deaths.

The theoretically reanimated consciousness then would, on their clocks, immediately reawaken and be able to 'watch' or 'hear' messages from their loved ones, left after their own biological demise. Aside from this informing the reanimated, there would be a great therapeutic benefit for the bereaved. Talking with the deceased is something that is already often done internally or symbolically at the resting place for the decomposing and decomposed deceased. Surely a service to leave 'Skype-like' messages for all involved with even a slightly better chance of being received would be a good thing.

This is clearly a major claim that many will instinctively resist. This book has taken what we collectively see and understand and withheld the temptation to look at any specific part of us as 'special', and so form a prejudiced assumption, and by logical projection, really, we arrive at a point where to lose the information that creates us is to allow us to die irretrievably. Yet, by taking a sample of DNA and video filing as much as possible (which we could all do now), plus improving brain scans, plus cryonics, brain preservation, or whatever else we come up with over coming decades (as well as investing in extending the survival of the current biological seat of this most precious illusion), we really have no reason to assume we cannot continue either now or after a subjectively unnoticed period of time, and we can challenge the finality of biological death.

It would seem likely that there are two key reasons why such vast numbers of people still adhere to what seem very much like fairytales in the forms of all of the world's religions: firstly, the fact that their loved ones enthusiastically promote these ideas to them when they are very young and they then conflate familial love with religious belief; and secondly, because the alternative of reconciling ourselves with interminable loss of loved ones is just unbearable. So we create and adhere to pretty ridiculous fairytales—surely what is proposed above is much less senseless. I for one hope that organisations that can last in perpetuity and are trustworthy can begin to store individuals' informatics now. Of course there are books to be written on how we should organise such things and the worlds we can create but:

Chapter Nine: Beyond the Subjectivity Trap

This is the ultimate reality of appreciating our lack of specialness and the truth that there isn't someone or something 'out there' deciding what we can or can't do. We are in no way special, but because we can now communicate, cooperate, and record our growing knowledge, our potential is clearly limitless. We are nothing… and… everything.